BOOK OF DATA

KIMBOLTON SCHOOL,
HUNTINGDON.

This book is the property of the school. It is issued on loan to the pupil whose name appears last on the list below, and loss or damage must be paid for. Names of previous users must not be obliterated.

Date	Name	Condition
Sept 01	Matthew Blake	~~B~~ New
4/09/03	R. Wilkinson	A
13/9/05	S. Roberts	A

A - *New* B - *Good* C - *Average* D - *Below Average*

Revised Nuffield Advanced Science

Editor of the first edition
R. D. Harrison

Editor of the revised edition
Hendrina Ellis

Contributing editors
Hendrina Ellis
R. D. Harrison
H. D. B. Jenkins

Contributors
Professor A. W. Brierley
B. E. Dawson
A. W. L. Dudeney
John Harris
B. J. Stokes
M. D. W. Vokins
E. J. Wenham

BOOK OF DATA

Revised Nuffield Advanced Science
Published for the Nuffield-Chelsea Curriculum Trust
by Longman Group Limited

Addison Wesley Longman Limited
Edinburgh Gate, Harlow, Essex CM20 2JE, England
and Associated Companies throughout the world

First published 1972
Revised edition first published 1984
Sixteenth impression 1998
Copyright © 1972, 1984, The Nuffield-Chelsea Curriculum Trust

Design and art direction by Ivan Dodd
Illustrations by Oxford Illustrators Limited

Filmset in Monophoto Times and Univers
Produced by Addison Wesley Longman Singapore Pte Ltd
Printed in Singapore

ISBN 0 582 35448 X

The Publisher's policy is to use paper manufactured from sustainable forests.

Cover photograph
Microchip Networks
Copyright Paul Brierley

Contents

Contents

Foreword

In planning the original Nuffield Advanced Science courses nearly twenty years ago it was decided that it would be extremely useful for students using our Chemistry, Physics, or Physical Science materials to be able to draw on a single *Book of data*. The first edition of this book was an immediate success, but only when we came to consider its revision did we realize that, in publishing it, we had performed a service not only for students taking our special Nuffield examinations but for those following other courses both in schools and in higher education. This wider use of the *Book of data* is something of which we have tried to take account in planning its revision.

We owe a great debt to Roger Harrison of the Open University, who as editor compiled the first edition of the *Book of data*. He has been most generous in the help he has given in the revision. He was assisted in the first edition by Hendrina Ellis whose services we have been most fortunate to secure in bringing out the revised edition. We are very grateful to her for her tireless work over the past three years in bringing the contents up to date with the most recent figures and researches and in devising new tables which add greatly to the value of the book. She has been advised in this work by the General Editors of the Revised Nuffield Advanced Chemistry and Physics materials respectively, B. J. Stokes and John Harris, and by Professor Ernest Coulson, organizer of the first edition of the Nuffield Advanced Chemistry project. She has also received notable help from Dr H. D. B. Jenkins of the University of Warwick and Dr Joe Lee of UMIST.

I would also like to acknowledge the work of William Anderson, publications manager to the Trust, his colleagues, and our publishers, the Longman Group, for their assistance in the publication of this book. The editorial and publishing skills they contribute are essential to effective curriculum development.

K. W. Keohane
Chairman, Nuffield-Chelsea Curriculum Trust

Introduction

From the Introduction to the first edition

I have long felt a need for an up-to-date compilation of basic physical data in SI units for use by my students, and had already begun to consider producing one when it was decided that there should be a book of data for the Nuffield Advanced courses in Physics, Physical Science, and Chemistry. I was therefore delighted when I was asked to undertake the task of compiling and editing this work: it has been a labour of love.

I am grateful to the Newcastle-upon-Tyne Polytechnic for permitting me to undertake this work partly during working hours and to use the Polytechnic computer and other facilities.

It is a pleasure to acknowledge the help received in preparing possible tables from various outside consultants, friends, colleagues, students, technicians, and the Nuffield headquarters teams, among them B. Britton, T. Burns, R. Day, A. E. Dodd (of the British Ceramic Research Association), P. J. Doyle (of the British Glass Industry Research Association), Sister D. Furtado, R. F. Hearman (of the Forest Products Research Laboratory), E. Henshall, R. Hodgson, D. J. Hucknall, J. A. Hunter, L. V. Kite, T. R. Manley, G. G. Matthews, T. Priest (of Exeter University), P. Pullar-Strecker (of C.I.R.I.A.), P. Roe, I. S. Simpson, J. Thompson, B. Tunnard, R. W. Tyler, and Miss C. A. Wigglesworth. My special thanks are due to Professor M. L. McGlashan, who went to enormous trouble to advise on the correct use of the new conventions. The remaining faults are due to me alone. Mrs E. Hadwin typed most of the script and Miss M. Nicholson punched the computer cards.

R. D. Harrison
The Open University
1971

Introduction to the revised edition

Those who have used the first edition of the *Book of data* will know how extensively this revised edition is based on its predecessor. However thirteen years of use and criticism, changes in syllabus, and availability of better sources of data, have resulted in many changes.

The content has changed. Conversion tables and tables of mathematical functions are no longer necessary now that all students use calculators. New tables on infra-red and nuclear magnetic resonance spectroscopy, on crystal structures and systems, and the shapes of molecules and ions, have been added in their place. Some tables have been extended to cover more items and categories of information; others have been pruned because of changes in emphasis in the curriculum.

The tables of physical and thermochemical properties of elements, inorganic compounds, and organic compounds, have been revised and extended by Dr H. D. B. Jenkins of the University of Warwick using the American National Bureau of Standards tables of data which were only partly available when the first edition was produced. These tables in section 5 make available an important source of information converted to SI units which should be most valuable to Chemistry students. The table of lattice energies is calculated from information in this edition, and the heats of combustion of organic compounds have been recalculated.

The design of the book has changed. Larger page size allows larger type size; a new system of sections and table numbers should allow students to find tables more easily, as should the enlarged Contents list; and the index is improved and extended.

I would particularly like to thank Roger Harrison, the editor of the first edition, for his help in revising tables in sections 1 to 3; Don Jenkins, for the massive task of providing the new data for section 5, and for his advice and new data in other sections; Dr J. Lee of UMIST who acted as external referee and meticulously corrected the use of symbols and conventions as well as providing new data; and Dr Jean Macqueen who provided the indexes. I would also like to thank Bryan Stokes, Organizer of Revised Nuffield Advanced Chemistry, and John Harris, Organizer of Revised Nuffield Advanced Physics, for their help and advice; and Professor Brierley, Bernard Dawson, Bill Dudeney, Michael Vokins, and Ted Wenham for providing data for the new tables.

Hendrina Ellis
Nuffield-Chelsea Curriculum Trust
1984

How to use this book

This book is divided into eight sections given on the first page of the Contents list; to help you find each section there is a black tab down the side of the book indicating its position.

The second and third pages of the Contents list contain all the table titles so if you know which section you want you can look through the relevant titles to find your table.

Until you are used to using this book, however, it is probably quickest to use one of the indexes. There is a *substance index*, so that you can quickly find information on a particular substance in which you are interested; and there is a *general index*, in which you can find references to properties, such as density or electronic structure (rather than to substances), and which also lists SI and non-SI units, symbols, physical constants, equations and formulae, etc.

When you find the entry you want, check the *definition* of the property at the start of the table (it might, for instance, relate to a different temperature to the one you are interested in); and check the *units* used at the head of the column.

The International System of Units (Système International d'Unités), which is abbreviated SI, is a coherent system of seven base units, two supplementary units, and an unlimited number of derived units (eighteen of which have special names). The base, supplementary, and specially named derived units are given below.

SI should be used in all branches of science, technology, industry, and commerce, as well as in everyday life, and is recognized throughout the world.

The great advantages of SI over some earlier systems of measurement are that there is a single unit for each quantity; there is a single set of electromagnetic units (and associated quantities); it is completely coherent; and used with unit prefixes (listed below) is completely decimal. These features greatly simplify calculations.

By coherent, we mean that all the derived units are formed by simple multiplication of the base or other derived units, without introduction of any numerical factor, even a power of ten. In consequence, when measurements expressed in base or derived units of SI are substituted in an equation the result will automatically be in the appropriate base or derived unit of SI.

BASE UNITS

These units and their associated quantities are dimensionally independent.

metre The unit of length is equal to the length of the path travelled by light in vacuum during a time interval of $1/299\,792\,458$ of a second.[A]
Unit symbol: **m**

kilogram The unit of mass is equal to the mass of the international prototype kilogram (a platinum-iridium cylinder) kept at the Bureau International des Poids et Mesures (BIPM), Sèvres, Paris.[B] *Unit symbol:* **kg**

second The unit of time is the duration of exactly $9\,192\,631\,770$ periods of the radiation corresponding to the transition between the two hyperfine levels of the ground state of the caesium-133 atom. *Unit symbol:* **s**

ampere The unit of electric current is that constant current which, if maintained in two straight parallel conductors, of infinite length and negligible cross-section, placed 1 metre apart in a vacuum, would produce a force between these conductors equal to 2×10^{-7} newton per metre of length. *Unit symbol:* **A**

kelvin The unit of thermodynamic temperature[C] is the fraction $1/273.16$ (exactly) of the thermodynamic temperature at the triple point of water.
Unit symbol: **K**

mole The unit of amount of substance is the amount of substance which contains as many elementary entities as there are atoms in 0.012 kilogram of pure carbon-12. *Note:* when the mole is used, the elementary entities must be specified and may be atoms, molecules, ions, electrons, other particles, or specified groups of such particles.
Unit symbol: **mol**

candela The unit of luminous intensity is the luminous intensity, in a given direction, of a source that emits monochromatic radiation of frequency 540×10^{12} hertz and that has a radiant intensity in that direction of $1/683$ watt per steradian.
Unit symbol: **cd**

SUPPLEMENTARY UNITS[D]

radian The unit of angle is the angle subtended at the centre of a circle by an arc of the circumference equal in length to the radius of the circle.
Unit symbol: **rad**

steradian The unit of solid angle is the solid angle subtended at the centre of a sphere of radius r by a portion of the surface of the sphere having an area r^2.
Unit symbol: **sr**

[A] This definition (1983) replaces the earlier definition of the metre as equal to $1\,650\,763.73$ wavelengths in vacuum of the radiation corresponding to the transition $2p_{10}-5d_5$ of the krypton-86 atom.
[B] A redefinition of the kilogram in terms of atomic constants will be adopted when the experimental precision warrants this.
[C] The International Unit of Pure and Applied Chemistry (IUPAC) accepts 'absolute temperature' as an equivalent name.

[D] At the choice of the user, these units may be regarded either as base or specially named derived units. It has been recommended by the International Committee on Weights and Measures (CIPM) that these units should be a 'class of dimensionless derived units', that is, they may be equated to unity.

DERIVED UNITS WITH SPECIAL NAMES

Frequency	**hertz**[A]	$Hz = s^{-1}$
Force	**newton**	$N = kg\,m\,s^{-2}$
Energy	**joule**	$J = kg\,m^2\,s^{-2}$
Power	**watt**	$W = kg\,m^2\,s^{-3}$
Pressure	**pascal**	$Pa = kg\,m^{-1}\,s^{-2}$
Luminous flux	**lumen**	$lm = cd\,sr$
Illuminance	**lux**	$lx = cd\,sr\,m^{-2}$
Electric charge	**coulomb**	$C = A\,s$
Electric potential difference	**volt**	$V = kg\,m^2\,s^{-3}\,A^{-1}$
Electric resistance	**ohm**	$\Omega = kg\,m^2\,s^{-3}\,A^{-2}$
Electric capacitance	**farad**	$F = A^2\,s^4\,kg^{-1}\,m^{-2}$
Magnetic flux	**weber**	$Wb = kg\,m^2\,s^{-2}\,A^{-1}$
Magnetic flux density	**tesla**	$T = kg\,s^{-2}\,A^{-1}$
Inductance	**henry**	$H = kg\,m^2\,s^{-2}\,A^{-2}$
Electric conductance	**siemens**	$S = \Omega^{-1} = kg^{-1}\,m^{-2}\,s^3\,A^2$
Activity (of a radioactive source; nuclear transformations per unit time)	**becquerel**	$Bq = s^{-1}$
Absorbed dose (of ionizing radiation)	**gray**	$Gy = J\,kg^{-1}$
Dose equivalent (of ionizing radiation)	**sievert**	$Sv = J\,kg^{-1}$

SI PREFIXES

Multiplying prefixes may be used with any unit symbols to indicate decimal multiples or fractions.

10^{18}	**exa**	**E**	10^{-1}	**deci**	**d**
10^{15}	**peta**	**P**	10^{-2}	**centi**	**c**
10^{12}	**tera**	**T**	10^{-3}	**milli**	**m**
10^{9}	**giga**	**G**	10^{-6}	**micro**	**μ**
10^{6}	**mega**	**M**	10^{-9}	**nano**	**n**
10^{3}	**kilo**	**k**	10^{-12}	**pico**	**p**
10^{2}	**hecto**	**h**	10^{-15}	**femto**	**f**
10	**deca**	**da**	10^{-18}	**atto**	**a**

It is strongly recommended that you write out the prefix as a power of ten before substituting a value into an equation.

The prefix and the unit symbol together form a single algebraic symbol. Thus km means 1000 m and km^{-2} means $(1000\,m)^{-2}$, or $10^{-6}\,m^{-2}$. Compound prefixes should not be used (for example, use nm but not mμm for $10^{-9}\,m$). In the case of the kilogram, SI prefixes should be attached to the gram g, not to the kilogram kg, in spite of the fact that the kilogram is the base unit; for example, mg not μkg for $10^{-6}\,kg$.

It is usually convenient to choose the prefix so that the number multiplying the unit (the measure) will lie between 0.1 and 999.9.

References: ASE (1979), ASE (1981), BS5555, BS5775, IUPAC, McGlashan, Royal Society (1975).

[A] To be used only when referring to periodic phenomena and not, for instance, for radioactive count rates.

The recommended symbols for a number of physical quantities are given on pages 6–9, together with the appropriate SI unit. SI units, definitions, and symbols are given in Table 1.1. Mathematical symbols are given in Table 1.3. Fundamental constants are given in Table 1.6.

Because there is no compulsion on the choice of symbols (and because the same letter may be recommended for several different quantities), a physical quantity symbol should always be defined precisely the first time it is used.

The following notes explain some of the conventions followed when using physical quantities, symbols for physical quantities, units, and symbols for units.

1 Physical quantity = numerical value × unit

This definition of a physical quantity as the product of a numerical value (sometimes called the measure) and a unit leads to the following conventions for labelling table headings and graph axes, and for substituting in formulae. The equation can be rearranged:

$$\frac{\text{physical quantity}}{\text{unit}} = \text{numerical value}$$

Tables of data give numerical values for various physical quantities. They must therefore have column headings in the form (physical quantity/unit), such as:

$$r_{\text{m}}/\text{nm} \qquad \text{and} \qquad \frac{\Delta H_{\text{f}}^{\ominus}}{\text{kJ mol}^{-1}}$$

Similarly graphs plot numerical values of physical quantities and must have their axes labelled (physical quantity/unit). This also means that graphical gradients, intercepts, and areas are numbers and must be interpreted accordingly.

When using formulae and equations (see Table 8.1) remember that the symbol for a physical quantity represents (numerical value × unit) and substitute *both* of these in the formula, for example:

Given that $F = ma$ (where F = force on particle, m = mass of particle, a = acceleration)
and that $m = 4\,\text{kg}$ and $a = 3\,\text{m s}^{-2}$
then $F = (4\,\text{kg})(3\,\text{m s}^{-2}) = 4 \times 3\,\text{kg m s}^{-2} = 12\,\text{kg m s}^{-2} = 12\,\text{N}$

2 Writing and printing of symbols for physical quantities and for units

Symbols for physical quantities are printed in italic (sloping) type, to distinguish them from symbols for units which are printed in roman (upright) type. It is not possible to make this distinction in handwriting and so confusion can arise, for example with V for potential difference and V for volt. In this case when symbols for units and symbols for physical quantities are used close together (such as V/V in a table heading) it may be advisable to spell out the unit symbol (V/volt).

Symbols for units are printed in roman (upright) type. A space should be left between the numerical value and the unit, for example 5 cm. In a compound unit spaces should be left between symbols, for example, $2\,\text{J mol}^{-1}\,\text{K}^{-1}$. SI prefixes should be printed or written close up to the symbol, for example, kJ, μmol. Unit symbols do not take the plural s, for example, 5 cm (not 5 cms).

3 Modifying signs, superscripts, and subscripts for use with symbols for physical quantities

A symbol for a physical quantity can be modified in various ways. For example:

	Mass of Earth	**Mass of Moon**
Capital and lower case letters	M	m
Subscript numbers	m_1	m_2
Subscript letters	m_{E}	m_{M}
Parentheses ( )	$m(\text{Earth})$	$m(\text{Moon})$

Experimental conditions can be put inside parentheses, for example, $n(H_2O(l), 300\,K, 590\,\mu m)$ for the refractive index of water at 300 K for wavelength 590 μm.

However, there are a number of superscripts and subscripts which can be used to indicate more briefly the physical state of a substance. (Be warned; there is some divergence between different authorities over these, and also usage has changed so that older books may use different conventions.)

Superscripts		Subscripts	
$°\,*\,\bullet$	pure substance	∞	limiting value at infinite dilution
id	ideal	g l s aq	gas, liquid, solid, aqueous states
$*$	excited electronic state	at	atomization
$\ddagger$	transition state, activated complex	c	combustion
$+\,-$	positive or negative ion or electrode	f	formation
$\ominus$	standard sign (see paragraph 4 below)	sub	sublimation
		tr	transition (usually solid state)
		m	melting b boiling

Square brackets around a physical quantity, for example [velocity] or [v], denote the dimension of the quantity. Square brackets round a chemical formula denote the concentration of the substance. Apart from these cases, brackets, parentheses, braces, etc, should not be used to modify a physical quantity.

The symbol Δ preceding a quantity symbol denotes the change of that quantity when a designated process (physical or chemical) occurs.

4 Standard sign $\ominus$

The standard sign is widely used in thermochemistry, but not always in the same way in different books. It is used for measurements made under (or calculated for) certain standard conditions of pressure, molality, etc. (and in some usages temperature), which should be stated the first time the symbol is used. (Where this is significant, correction for non-ideality is implicit in the use of the symbol.) The standard pressure is usually chosen to be 1 atm (101 325 Pa), the standard molality is usually $1\,mol\,kg^{-1}$, and the standard temperature where appropriate is often 298 K. In this book the standard sign is frequently used. For example, $\Delta H_f^{\ominus}$ is the standard molar enthalpy change of formation. In all cases the chosen standard pressure is 1 atm, and the experimental temperature is (with some exceptions) 298 K.

5 'Specific' and 'molar'

These words are used to qualify physical quantities in the following way.

'Molar' means 'divided by amount of substance'.*

For example, volume V molar volume $V/n = V_m$ (where n is number of moles present)
There are a few exceptions to this, for example, molar conductivity where 'molar' means 'divided by concentration'. 'Molar' does *not* mean 'of 1 mole of substance (or reaction)'.

'Specific' means 'divided by mass'.

For example, volume V specific volume $V/m = v$ (where m is the total mass)
The convention for symbols for physical quantities is:
Capital letters refer to total quantities, e.g. H, C.
Capital letters with subscript $_m$ refer to molar quantities, e.g. H_m, C_m. (But commonly $_m$ is omitted where it has already been stated that molar quantities are intended.)
Capital letters with subscript $_B$ (where B stands for a chemical species) represent the partial molar quantity in a mixture, e.g. H_{Al} for the partial molar enthalpy of aluminium in an alloy.
Lower case letters stand for specific quantities, e.g. h for specific enthalpy, c for specific heat capacity.

* Or for chemical reactions, 'divided by change of extent of reaction'.

	Quantity	Symbol	SI unit
1	absorption coefficient (acoustic)	α_a	—
2	absorption factor (radiation)	α	—
3	acceleration	a	m s^{-2}
4	acceleration of free fall	g	m s^{-2}
5	acceleration, angular	α	rad s^{-2}
6	activation energy	$E, E^{\ddagger}$	J mol^{-1}
7	activity (radioactive)	A	Bq, s^{-1}
8	activity coefficient of substance B	f_B, γ_B, y_B	—
9	amount of substance B	n_B	mol
10	amplitude	a or x_0 etc	as appropriate
11	angle	$\alpha, \beta, \gamma, \theta, \Phi$ etc	rad^A
12	angle of contact	θ	rad^A
13	angle of deviation	D	rad^A
14	angle of prism	A	rad^A
15	angle of optical rotation	α	rad^A
16	angular acceleration	α	rad s^{-2}
17	angular momentum	L, b, p_θ	J s
18	angular velocity	ω	rad s^{-1}
19	area	A, S	m^2
20	average speed	$\bar{c}, \bar{u}$	m s^{-1}
21	atomic number	Z	—
22	attenuation coefficient (particles)	μ	m^{-1}
23	Avogadro constant	L, N_A	mol^{-1}
24	Boltzmann constant	k	J K^{-1}
25	Bragg angle	θ	rad^A
26	breadth	b	m
27	bulk modulus	K	N m^{-2}
28	capacitance	C	F
29	charge density (surface)	σ	C m^{-2}
30	charge density (volume)	ρ	C m^{-3}
31	charge number of ion	z	—
32	coefficient of friction	μ	—
33	compressibility	κ	m^2 N^{-1}
34	concentration of substance B	$c_B, [B]$	mol m^{-3}
35	conductance	G	S $= \Omega^{-1}$
36	conductivity, electrical	σ, γ	S m^{-1}
37	conductivity, thermal	λ, k	W m^{-1} K^{-1}
38	coordinate (Cartesian)	x, y, z	m
39	cubic expansivity	γ, α	K^{-1}
40	decay constant, radioactive	λ	s^{-1}
41	degree of dissociation	α	—
42	density	ρ	kg m^{-3}
43	diameter	d	m
44	dispersive power	ω	—
45	distance along path	s, L	m
46	electric charge	Q	C
47	electric current	I	A
48	electric current density	J	A m^{-2}
49	electric dipole moment	p, μ	C m
50	electric displacement	D	C m^{-2}
51	electric field strength	E	N C^{-1}, V m^{-1}
52	electric flux	Ψ	C
53	electric flux density	D	C m^{-2}
54	electric polarization	P	C m^{-2}
55	electric potential	V, ϕ	V
56	electric potential difference	V, U	V
57	electric susceptibility	χ_e	—
58	electrolytic conductivity	κ	S m^{-1}
59	electromagnetic moment	m	A m^2
60	electromotive force	E	V
61	electron mass	m_e	kg
62	elementary (electron) charge	e	C
63	emissivity	ε	—
64	energy	E, W	J
65	energy, internal	U, E	J
66	energy, kinetic	E_k, T, K	J
67	energy, potential	E_p, V, Φ	J
68	energy, radiant	Q, Q_e	J
69	enthalpy	H	J mol^{-1}
70	entropy	S	J K^{-1} mol^{-1}

^{A} In practice, these will often be measured in degrees (°).

	Quantity	Symbol	SI unit
71	equilibrium constant	K	as appropriate
72	expansivity, cubic	γ, α	K^{-1}
73	expansivity, linear	α, λ	K^{-1}
74	Faraday constant	F	$C\,mol^{-1}$
75	field strength, electric	E	$V\,m^{-1}$
76	field strength, magnetic	H	$A\,m^{-1}$
77	focal length	f	m
78	force	F	N
79	frequency	ν, f	Hz, s^{-1}
80	frequency, angular (pulsatance)	ω	$rad\,s^{-1}$
81	frequency, rotational	n	s^{-1}
82	Gibbs free energy[A]	G	$J\,mol^{-1}$
83	gravitational constant	G	$N\,m^2\,kg^{-2}$
84	grating spacing or slit separation	d	m
85	half-life (radioactive or reaction)	$T_{\frac{1}{2}}, t_{\frac{1}{2}}$	s
86	Hall coefficient	R_H	$m^3\,C^{-1}$
87	heat capacity	C	$J\,K^{-1}$
88	heat flow rate	Φ	W
89	heat, quantity of	Q, q	J
90	height	h	m
91	Helmholtz free energy function	A, F	J
92	image distance	v	m
93	impedance	Z	Ω
94	impulse	p	$N\,s$
95	internal energy	U, E	J
96	ionic strength	I	$mol\,kg^{-1}$
97	kinetic energy	E_k, T, K	J
98	latent heat (enthalpy change)	$L, \Delta H$	J
99	length	l	m
100	light, quantity of	Q, Q_v	J or $lm\,s$
101	linear expansivity	α, λ	K^{-1}
102	luminance	L, L_v	$cd\,m^{-2}$
103	luminous flux	Φ, Φ_v	lm
104	luminous intensity	I, I_v	cd
105	magnetic field strength	H	$A\,m^{-1}$

	Quantity	Symbol	SI unit
106	magnetic flux	Φ	Wb
107	magnetic flux density	B	T
108	magnetic moment	m	$A\,m^2$
109	magnetic polarization	J	T
110	magnetic susceptibility	χ_m	—
111	magnetization	M	$A\,m^{-1}$
112	magnetomotive force	F_m	A
113	magnification, linear	m	—
114	magnifying power	M	—
115	mass	m	kg
116	mass excess	Δ	kg
117	mass number	A	—
118	mass of electron	m_e	kg
119	mass of neutron	m_n	kg
120	mass of proton	m_p	kg
121	mean free path	l, λ	m
122	mean life (radioactive)	τ	s
123	molality of substance B	m_B	$mol\,kg^{-1}$
124	molar conductivity (conductance)	Λ	$S\,m^2\,mol^{-1}$
125	molar volume	V_m	$m^3\,mol^{-1}$
126	molar mass of a compound B	M_B	$kg\,mol^{-1}$
127	molar mass of an element	A	$kg\,mol^{-1}$
128	molecular mass	m	kg
129	molecular velocity speed	$c(u, v, w)$	$m\,s^{-1}$
130	mole fraction of substance B	x_B, y_B	—
131	moment of couple	T	$N\,m$
132	moment of force	M	$N\,m$
133	moment of inertia	I, J	$kg\,m^2$
134	momentum	p	$N\,s$
135	mutual inductance	M, L_{12}	H
136	neutron number	N	—
137	nucleon number	A	—
138	number density of molecules	n	m^{-3}
139	number of molecules	N	—
140	number of turns on coil	N	—

[A] Otherwise known as Gibbs function or more recently, Gibbs energy. Nuffield Advanced Chemistry materials use the form given above.

#	Quantity	Symbol	SI unit
141	number of turns per unit length of coil	n	$\mathrm{m^{-1}}$
142	object distance	u	m
143	order of reflection or interference	n	—
144	osmotic pressure	Π	Pa
145	packing fraction	f	—
146	Peltier coefficient	Π	V
147	period	T	s
148	permeability (magnetic)	μ	$\mathrm{H\,m^{-1}}$
149	permeability of vacuum	μ_0	$\mathrm{H\,m^{-1}}$
150	permeability, relative	μ_r	—
151	permittivity	ε	$\mathrm{F\,m^{-1}}$
152	permittivity of vacuum	ε_0	$\mathrm{F\,m^{-1}}$
153	permittivity, relative	ε_r	—
154	phase angle	ϕ	rad (or °)
155	Planck constant	h	$\mathrm{J\,s}$
156	Planck constant divided by 2π	$\hbar$	$\mathrm{J\,s}$
157	Poisson ratio	μ, ν	—
158	potential energy	E_p, V, Φ	J
159	power	P	W
160	power factor	$\cos\phi$	—
161	power of lens	F	$\mathrm{rad\,m^{-1}}$
162	pressure	p, P	$\mathrm{Pa}\,(=\mathrm{N\,m^{-2}})$
163	pulsatance (angular frequency)	ω	$\mathrm{rad\,s^{-1}}$
164	quantum number (principal)	n	—
165	radius	r	m
166	radius of gyration	k	m
167	rate constant of $(n+1)^{\text{th}}$ order reaction	k, k_r	$\mathrm{m^{3n}\,mol^{-3n}\,s^{-1}}$
168	ratio $C_p/C_V = c_p/c_V$	γ	—
169	reactance	X	Ω
170	reflection coefficient (factor)	ρ	—
171	refractive index	n	—
172	relative atomic mass of an element (atomic weight)	A_r	—
173	relative molecular mass of a substance (molecular weight)	M_r	—
174	resistance	R	Ω
175	resistivity	ρ	$\Omega\,\mathrm{m}$
176	self inductance	L	H
177	shear modulus	G	$\mathrm{Pa}\,(=\mathrm{N\,m^{-2}})$
178	slit separation or grating spacing	d	m
179	solid angle	Ω, ω	sr
180	sound intensity	I, J	$\mathrm{W\,m^{-2}}$
181	specific charge (electron)	e/m_e	$\mathrm{C\,kg^{-1}}$
182	specific heat capacity	c	$\mathrm{J\,kg^{-1}\,K^{-1}}$
183	specific heat capacity at constant pressure	c_p	$\mathrm{J\,kg^{-1}\,K^{-1}}$
184	specific heat capacity at constant volume	c_v	$\mathrm{J\,kg^{-1}\,K^{-1}}$
185	speed	$u, v, w: c$	$\mathrm{m\,s^{-1}}$
186	speed of electromagnetic waves (light) in vacuum	c	$\mathrm{m\,s^{-1}}$
187	speed of sound	c	$\mathrm{m\,s^{-1}}$
188	Stefan–Boltzmann constant	σ	$\mathrm{W\,m^{-2}\,K^{-4}}$
189	strain, linear	ε, e	—
190	strain, shear	γ	—
191	strain, volume	θ	—
192	stress, normal (compressive)	σ	$\mathrm{Pa}\,(=\mathrm{N\,m^{-2}})$
193	stress, shear	τ	$\mathrm{Pa}\,(=\mathrm{N\,m^{-2}})$
194	stress, volume	p	$\mathrm{Pa}\,(=\mathrm{N\,m^{-2}})$
195	surface tension (energy)	γ, σ	$\mathrm{N\,m^{-1}}$ $(=\mathrm{J\,m^{-2}})$
196	temperature, common (Celsius)[A]	θ, t	°C
197	temperature, thermodynamic (absolute)	T	K
198	temperature difference	θ	K
199	thermal capacity	C	$\mathrm{J\,K^{-1}}$
200	thermal conductivity	λ, k	$\mathrm{W\,m^{-1}\,K^{-1}}$
201	thermoelectric power (differential)	S	$\mathrm{V\,K^{-1}}$
202	thickness	d, δ	m
203	Thomson coefficient	μ	$\mathrm{V\,K^{-1}}$

[A] By definition, common temperature is expressed in non-SI units.

	Quantity	Symbol	SI unit
204	time	t	s
205	time constant	τ	s
206	torque	T	N m
207	transmission coefficient (transmittance)	τ	—
208	transport number	t	—
209	van der Waals coefficients	a and b	$N\,m^4\,mol^{-2}$, $m^3\,mol^{-1}$
210	velocity	u, v, w	$m\,s^{-1}$
211	viscosity (dynamic)	η, μ	$N\,s\,m^{-2}$
212	viscosity (kinematic)	ν	$m^2\,s^{-1}$
213	volume	V	m^3
214	volume expansivity	γ, α	K^{-1}
215	wavelength	λ	m
216	wavenumber	$\sigma, \tilde{\nu}$	m^{-1}
217	weight	W, G, P	N
218	work	W	J
219	work function	ϕ	V
220	Young modulus	E	$N\,m^{-2}$

References: ASE (1979), ASE (1981), Royal Society (1975), McGlashan.

Symbol	Meaning
$=$	is equal to
$\neq$	is not equal to
$\equiv$	is identically equal to
$\triangleq$	corresponds to
$\approx$	is approximately equal to
$\propto$ (or $\sim$)[A]	is proportional to
$\rightarrow$	tends towards, approaches
$\Rightarrow$	implies that
$A/B, \dfrac{A}{B}, A\,B^{-1}$	A divided by B (/ is called a solidus)
a^n	a raised to power n
$a^{\frac{1}{2}}, \sqrt{a}, \sqrt{a}$	square root of a
$a^{1/n}, \sqrt[n]{a}$	n^{th} root of a
$\lim\limits_{x \to a} f(x)$	limit of $f(x)$ as $x \to a$
∞	infinity
Δx	finite increment of x
δx	infinitesimal increment of x; variation of x
$\dfrac{df}{dx}, df/dx, f'(x)$	differential coefficient of $f(x)$ wrt[B] x
$\dfrac{d^n f}{dx^n}, d^n f/dx^n, f^{(n)}(x)$	differential coefficient of $f(x)$ wrt x n times
$\dot{x}, dx/dt$	differential coefficient of x wrt t
$>, <$	is larger than, is less than
$\gg, \ll$	is much larger than, is much less than
$\geqslant, \geqq, \geq$	is larger than or equal to
$\leqslant, \leqq, \leq$	is less than or equal to
$\pm$	plus or minus
$\parallel$	is parallel to
$\perp$	is perpendicular to
$\angle A, \hat{A}$	angle A
$r!$	factorial $r = r(r-1)(r-2)\ldots \times 2 \times 1$
$\dbinom{n}{r}$	binomial coefficient $= \dfrac{n!}{r!(n-r)!}$
$\bar{x}, x_{\text{av}}, \langle x \rangle$	average value of x
$x_{\max}$ or $\hat{x}$	maximum value of x
$x_{\min}$ or $\check{x}$	minimum value of x
$x_{\text{r.m.s.}}$ (or x_{eff})	root mean square (r.m.s.) value of x
x_0	peak value of x
$\int f(x)\,dx$	indefinite integral of f wrt x
$\int_a^b f(x)\,dx$	definite integral of f wrt x
$\oint f(x)\,dx$	integral of $f(x)$ round a closed path
$\sum x_i$ or $\sum\limits_{i=1}^{n} x_i$	sum of members of the set $x_1, \ldots x_n$
$\prod x_i$ or $\prod\limits_{i=1}^{n} x_i$	product of members of the set $x_1, \ldots x_n$

(The brace after "is parallel to", "is perpendicular to", "angle A" reads:) mainly used in pure mathematics

[A] The symbol $\sim$ is commonly used in physical science for 'is of the order of magnitude'.

[B] wrt means 'with respect to'.

Symbol	Meaning
$\lvert x \rvert$	the modulus function or absolute value of x
$\lvert A \rvert$	determinant of the square set A_{ij}
e^x, $\exp x$	exponential function of x
e	base of natural logarithms $= 2.718\,281\,828\,5$ (to 10 decimal places)
π	ratio of circle circumference to diameter $= 3.141\,592\,653\,6$ (to 10 decimal places)
$\ln x$, $\log_e x$	natural logarithm of x
$\lg x$, $\log x$, $\log_{10} x$	logarithm to base 10 of x (common logarithm)
$\log_a x$	logarithm to base a of x
$\mathrm{lb}\, x$, $\log_2 x$	binary logarithm of x (to base 2)
$\sin x$, $\cos x$, $\tan x$, $\sec x$, $\operatorname{cosec} x$, $\cot x$	trigonometric functions (See Table 8.1, 'Mathematics formulae'.)
$\arcsin x$, or $\sin^{-1} x$, etc.	argument of trigonometric function

NAMES OF MODIFYING SIGNS

$^-$ bar	$\cdot$ dot	$_x$ subscript	( ) parentheses
† dagger	$\hat{\ }$ hat	^{x} superscript	[ ] brackets
$'$ dash (or prime)	* star (asterisk)	$\sim$ tilde	{ } braces

CONVENTIONS FOLLOWED IN WRITING NUMBERS AND MATHEMATICAL STATEMENTS

The *decimal point* is indicated by a dot on the line or, in continental texts, by a comma: e.g. 123.45 or 123,45. A zero should be placed before the point in numbers less than unity: e.g. 0.123 not .123.

Long numbers should be written in groups of three digits, with a space between groups: e.g. 1 234.567 89. Commas should *not* be used to separate thousands to avoid confusion with the continental decimal point.

Significant terminal zeros to the right of the decimal point should not be omitted, for example, $25.00\,\mathrm{cm}^3$ should not be shortened to $25\,\mathrm{cm}^3$ when it is known that the two zeros are significant. Significance (or precision) ambiguity for measures terminating in zeros and presented without a decimal fraction are best resolved by use of powers of ten, for example, $1.23 \times 10^3\,\mathrm{m}$ or $1.230 \times 10^3\,\mathrm{m}$ or $1.23\,\mathrm{km}$ or $1.230\,\mathrm{km}$ (as appropriate), is preferable to the ambiguous $1230\,\mathrm{m}$.

The *argument of a function* should be enclosed in brackets (except for standard functions with not more than two symbols in the argument). Thus $f(x)$, $\exp\{(\tau - \tau_0)/\lambda\}$ but e^{kx}, $\sin wt$.

Letter		Name	Letter		Name	Letter		Name
A	α	Alpha	I	ι	Iota	P	ρ	Rho
B	β	Beta	K	κ	Kappa	Σ	σ	Sigma
Γ	γ	Gamma	Λ	λ	Lambda	T	τ	Tau
Δ	δ	Delta	M	μ	Mu	Υ	υ	Upsilon
E	$\varepsilon\,\epsilon$	Epsilon	N	ν	Nu	Φ	$\phi\,\varphi$	Phi
Z	ζ	Zeta	Ξ	ξ	Xi	X	χ	Chi
H	η	Eta	O	o	Omicron	Ψ	ψ	Psi
Θ	$\theta\,\vartheta$	Theta	Π	π	Pi	Ω	ω	Omega

References for Tables 1.3 and 1.4: McGlashan, Royal Society (1975).

Many non-SI units are now defined exactly in terms of SI; those marked * can only be related to SI units via fundamental constants and the relationship is therefore restricted by the precision to which the constants are known. (Units not *definable* in terms of SI units are not considered satisfactory units.) Exact values are printed in bold type. Multiples and submultiples of some of the non-SI units may be formed, where appropriate, by prefixes as in the SI, provided that this can be done without ambiguity (*not* min for 0.001 in, etc). Names of units within the SI are underlined. (This table may also be used to find out the meaning of unfamiliar units and unit symbols.)

Students are advised to convert all non-SI units and prefixed SI units to base (or specially named derived) SI units before carrying out calculations. Compound units not given in this list may be converted by the use of the appropriate conversion factor for each factor of the unit symbol, for example,

$$1 \, lb \, ft^{-3} = \{0.4536 \, (kg \, lb^{-1})/(0.3048)^3 \, (m \, ft^{-1})^3\} \, lb \, ft^{-3} = 16.0 \, kg \, m^{-3}.$$

The relationship between CGS and SI electromagnetic units in these tables is that of correspondence ($\hat{=}$) and not of equality, because SI electrical units are based on four dimensionally independent base units (m, kg, s, A) and CGS on only three (cm, g, s). It is incorrect to write, for example, 1 e.m.u. of current = 10 A because these are formally units of dimensionally different (albeit proportional) quantities.

Unit	Symbol	SI equivalent		
acre (= 4840 yd^2)	acre	4.047	$\times 10^3$	m^2
ångström	Å		**10^{-10}**	**m**
astronomical unit (Earth–Sun)*	AU	1.496	$\times 10^{11}$	m
atmosphere	atm	**101 325**		Pa
atomic mass unit (unified)*	u	1.661	$\times 10^{-27}$	kg
bar	bar		**10^5**	**Pa**
barn (unit of nuclear area)	b		**10^{-28}**	**m^2**
becquerel (SI: activity (of a radioactive source))	Bq		**1**	**s^{-1}**
biot ($\hat{=}$CGS: e.m.u. current or abampere)^A	Bi		**10**	**A**
British thermal unit	Btu	1.055	$\times 10^3$	J
	Btu h^{-1}	2.931	$\times 10^{-1}$	W
bushel (U.K.) (= 8 gal)	—	3.637	$\times 10^{-2}$	m^3
calorie (thermochemical)	cal$_{th}$	**4.184**		**J**
coulomb (SI: electric charge)	C		**1**	**A s**
cubic foot	ft^3	2.832	$\times 10^{-2}$	m^3
cubic yard	yd^3	0.7646		m^3
curie (radioactivity)	Ci	**3.7**	**$\times 10^{10}$**	**Bq**
day	d	**8.6400**	**$\times 10^4$**	**s**
debye*	D	3.336	$\times 10^{-30}$	C m
decibel^B	dB			
degree (angular)	...°	$\pi/180$ (1.745 $\times 10^{-2}$)		rad
degree Celsius^C	°C		**1**	**K**
degree Fahrenheit^C	°F	5/9 (0.5556)		K

* See introduction to the table.

^{A} The biot and the e.m.u. of current correspond to the same physical situation but whereas the latter can be expressed in terms of the three CGS base units, the former is a fourth base unit in an extension of CGS. Because biot and ampere are dimensionally equivalent, *direct* relationship (rather than correspondence) exists.

^{B} The decibel is used to express $10 \log_{10} P/P^{\ominus}$, where P is a power and $P^{\ominus}$ a standard power which must be specified. In acoustics,

$P^{\ominus}$ is generally 10^{-2} W. Sound intensities I are generally expressed in dB as $10 \log_{10} I/I^{\ominus}$, with $I^{\ominus} = 10^{-2}$ W m^{-2}. In electrical engineering, amplification ratios are expressed in dB, and e.m.f.s are given in dB as $20 \log_{10} V/V^{\ominus}$, where $V^{\ominus}$ is often 10^{-6} V. The decibel is not strictly a unit in the sense of the other units in this table but it is very useful for measurements which range over many powers of 10.

^{C} Temperatures convert according to the formulae:
$$t_C/°C = (5/9)[(t_F/°F) - 32] = T/K - 273.15.$$

Unit	Symbol	SI equivalent		
dioptre (unit of power of lens)	D, dpt		**1**	**m^{-1}**
dyne (CGS: force)	dyn		**10^{-5}**	**N**
electronvolt*	eV	1.602	$\times 10^{-19}$	**J**
erg (CGS: energy)	erg		**10^{-7}**	**J**
farad (SI: capacitance)	F		**1**	**C V^{-1}**
foot	ft	**0.3048**		**m**
foot pound-force	ft lbf	1.356		N m (or J)
franklin ($\triangleq$CGS: e.s.u. charge or statcoulomb)[A]	Fr	3.336	$\times 10^{-10}$	C
gallon (UK)	gal (UK)	4.546	$\times 10^{-3}$	**m^3**
gauss (CGS: e.m.u. flux density)	G	$\triangleq 1.000$	$\times 10^{-4}$	**T**
gray (SI: absorbed dose (of ionizing radiation))	Gy		**1**	**J kg^{-1}**
hectare (land area)	ha		**10^4**	**m^2**
henry (SI: inductance)	H		**1**	**J A^{-2}**
hertz (SI: frequency)	Hz		**1**	**s^{-1}**
hour	h	**3600**		s
horsepower	hp	745.7		W
horsepower hour	hp h	2.685	$\times 10^6$	J
hundredweight (long, UK)	cwt	50.80		kg
inch	in	**2.54**	$\times 10^{-2}$	**m**
joule (SI: energy)	J		**1**	**N m**
kilogram-force (kilopond)	kgf (kp)	9.807		N
kilowatt hour	k Wh	**3.6**	$\times 10^6$	**J**
knot (international)	kn	0.5144		m s^{-1}
light year*		9.461	$\times 10^{15}$	m
litre[B]	l, L		**10^{-3}**	**m^3**
litre atmosphere	1 atm, L atm	**101.325**		**J**
lambert	L	3.183	$\times 10^4$	cd m^{-2}
lumen (SI: luminous flux)	lm		**1**	**cd sr**
lux (SI: illumination)	lx		**1**	**lm m^{-2}**
maxwell (CGS: e.m.u. magnetic flux)	Mx		**10^{-8}**	**Wb**
mho (reciprocal ohm)	mho		**1**	**S ($=\Omega^{-1}$)**
micron	μ		**10^{-6}**	**m**
mile (nautical)	n mile	**1.852**	$\times 10^3$	**m**
mile (statute)	mile	1.609	$\times 10^3$	m
mile per hour	mile h^{-1}	0.4470		m s^{-1}
millimetre of mercury	mm Hg	133.3		Pa
minute (angle)	…$'$	2.909	$\times 10^{-4}$	rad
minute (time)	min	**60**		s
newton (SI: force)	N		**1**	**kg m s^{-2}**
oersted (CGS: e.m.u. magnetic field)	Oe	$\triangleq 1000/4\pi$ (79.58)		A m^{-1}
		$\triangleq 1.000$	$\times 10^{-4}$	T (in vacuo)

* See introduction to the Table.

[A] The franklin and the e.s.u. of charge correspond to the same physical situation, but whereas the latter can be expressed in terms of the three CGS base units, the former is a fourth base unit in an extension of CGS. Because franklin and coulomb are dimensionally equivalent *direct* relationship (rather than correspondence) exists.

[B] The definition of the litre as 1000.028 cm^3 was abandoned in 1964. Because of the possibility of ambiguity the litre should not be used for high precision work: dm^3 is unambiguous. Furthermore the lower case symbol for litre can often be confused with unity.

Unit	Symbol	SI equivalent		
ohm (SI: resistance)	Ω		**1**	**VA^{-1}**
ounce (avoirdupois)	oz	2.835	$\times 10^{-2}$	kg
ounce (fluid UK)	fl oz (UK)	2.841	$\times 10^{-5}$	m^3
parsec*	pc	3.086	$\times 10^{16}$	m
pascal (SI: pressure)	Pa		**1**	**$N\,m^{-2}$**
phon (loudness level)[A]	phon			
pint (UK)	pt (UK)	5.682	$\times 10^{-4}$	m^3
poise (CGS: dynamic viscosity)	P		**10^{-1}**	**$kg\,m^{-1}\,s^{-1}$**
pound	lb	**0.453 592 37**		**kg**
pound-force	lbf	4.448		N
pound-force foot	lbf ft	1.35ᴏ		N m
pound-force per sq. inch[B]	lbf in^{-2}	6.895	$\times 10^3$	$N\,m^{-2}$
pound-force per sq. foot	lbf ft^{-2}	47.88		$N\,m^{-2}$
rad	rad		10^{-2}	Gy
rem[C]	rem		10^{-2}	Sv
röntgen	R	2.58	$\times 10^{-4}$	$C\,kg^{-1}$
second (angle)	...″	4.848	$\times 10^{-6}$	rad
siemens (SI: reciprocal ohm)	S		**1**	**$\Omega^{-1} (=A\,V^{-1})$**
sievert (SI: dose equivalent (of ionizing radiation))	Sv		**1**	**$J\,kg^{-1}$**
square foot	ft^2	9.290	$\times 10^{-2}$	m^2
square inch	in^2	6.452	$\times 10^{-4}$	m^2
square mile	mile^2	2.590	$\times 10^6$	m^2
square yard	yd^2	0.8361		m^2
standard atmosphere	atm	**101 325**		**Pa**
statvolt (CGS: e.s.u. electric potential)	—	$\cong 2.998$	$\times 10^2$	V
stokes (CGS: kinematic viscosity)	St		**10^{-4}**	**$m^2\,s^{-1}$**
talbot	—		**1**	**lm s**
tesla (SI: magnetic flux density)	T		**1**	**$Wb\,m^{-2}$** **$(=V\,s\,m^{-2})$**
therm (100 000 Btu)	therm	1.055	$\times 10^8$	J
ton (UK long, 2240 lb)	ton	1.016	$\times 10^3$	kg
ton-force	ton f	9.964	$\times 10^3$	N
ton-force per square inch	tonf in^{-2}	1.544	$\times 10^7$	$N\,m^{-2}$
tonne (metric ton)[D]	t		**10^3**	**kg**
torr (=mmHg to 1 in 10^7)	Torr	133.3		Pa
volt (SI: electric potential difference)	V		**1**	**$J\,C^{-1}$**
watt (SI: power)	W		**1**	**$J\,s^{-1}$**
weber (SI: magnetic flux)	Wb		**1**	**$V\,s\,(J\,s\,C^{-1})$**
X unit (approx. 0.001 Å	Xu	1.002	$\times 10^{-13}$	m
yard	yd	0.9144		m
year (tropical)*	a	3.156	$\times 10^7$	s

* See introduction to the table.

[A] The phon is a subjective (that is, depends on the individual) unit of loudness on a decibel scale.

[B] The abbreviation psi is often used, but is not recommended.

[C] 1 rem of any ionizing radiation produces the same biological effect in human beings as 1 R of X-rays.

[D] Tonne is used for commercial and engineering purposes but is not SI. Mg is preferable for scientific work.

Reference: Weast.

Fundamental constants are quoted from CODATA Bulletin No. 11 and are internationally recommended values. Errors are given in brackets and appertain to the last two digits. D denotes an exact value by definition. Quantities named in bold type were determined directly. Other quantities were obtained from these and other experiments by calculation. For values of π and e see Table 1.3 'Mathematical symbols'.

Quantity	Symbol	Value	Unit
FUNDAMENTAL CONSTANTS			
Speed of light in vacuum	c	299 792 458 (D)	$m\,s^{-1}$
permeability of vacuum[A]	μ_0	$4\pi \times 10^{-7} =$ 12.566 370 614 4 $\times 10^{-7}$ (D)	$H\,m^{-1}$
permittivity of vacuum[A]	ε_0	8.854 187 82(7) $\times 10^{-12}$	$F\,m^{-1}$
Faraday constant	F	9.648 456(27) $\times 10^{4}$	$C\,mol^{-1}$
Avogadro constant	L, N_A	6.022 045(31) $\times 10^{23}$	mol^{-1}
unified atomic mass unit	m_u	1.660 565 5(86) $\times 10^{-27}$	kg
Boltzmann constant	k	1.380 662(44) $\times 10^{-23}$	$J\,K^{-1}$
elementary charge	e	1.602 189 2(46) $\times 10^{-19}$	C
rest mass of electron	m_e	0.910 953 4(47) $\times 10^{-30}$	kg
charge: mass ratio of electron	e/m_e	1.758 804 7(49) $\times 10^{11}$	$C\,kg^{-1}$
mass of proton	m_p	1.672 648 5(86) $\times 10^{-27}$	kg
mass ratio proton: electron	m_p/m_e	1836.151 52(70)	
rest mass of neutron	m_n	1.674 954 3(86) $\times 10^{-27}$	kg
rest mass of hydrogen atom	m_H	1.673 559 5(86) $\times 10^{-27}$	kg
Rydberg constant	R_∞	1.097 373 177(83) $\times 10^{7}$	m^{-1}
Rydberg constant (hydrogen)	R_H	1.096 775 78(11) $\times 10^{7}$	m^{-1}
Planck constant	h	6.626 176(36) $\times 10^{-34}$	$J\,Hz^{-1}$
	$h/2\pi = \hbar$	1.054 588 7(57) $\times 10^{-34}$	J s
Bohr magneton	$\mu_B = e\hbar/2m_e$	9.274 078(36) $\times 10^{-24}$	$J\,T^{-1}$
Nuclear magneton	$\mu_N = e\hbar/2m_p$	5.050 824(20) $\times 10^{-27}$	$J\,T^{-1}$
Bohr radius	a_0	0.529 177 06(44) $\times 10^{-10}$	m
gravitational constant	G	6.6720(41) $\times 10^{-11}$	$N\,m^2\,kg^{-2}$
standard gravity[B]	g_n	9.806 65 (D)	$m\,s^{-2}$ or $N\,kg^{-1}$
standard density of mercury[B]	$\rho(Hg)$	1.359 51 $\times 10^{4}$	$kg\,m^{-3}$
molar volume of ideal gas[C]	V_m	0.022 413 83(70)	$m^3\,mol^{-1}$
gas constant	R	8.314 41(26)	$J\,K^{-1}\,mol^{-1}$
speed of sound in air	$c(273\,K)$	3.313 6 $\times 10^{2}$	$m\,s^{-1}$
OTHER CONSTANTS			
radius of Earth[D]	$r_\oplus$ or r_E	6.370 949 $\times 10^{6}$	m
mass of Earth	$m_\oplus$ or m_E	5.976 3 $\times 10^{24}$	kg
mean distance Earth to Sun[D]	AU or A	1.495 99 $\times 10^{11}$	m
solar constant[E]		1.40 (± 0.03) $\times 10^{3}$	$W\,m^{-2}$
Earth's horizontal magnetic field[F]	H	1.87 $\times 10^{-5}$	T
Earth's vertical magnetic field[F]	Z	4.36 $\times 10^{-5}$	T
Earth's equivalent dipole	m	8.1 $\times 10^{22}$	$A\,m^2$
maximum density of water	$\rho(H_2O, 277.13\,K)$	0.999 973 $\times 10^{3}$	$kg\,m^{-3}$
temperature of 'ice point'	T_{ice}	273.150 0(1)	K

[A] $\mu_0\varepsilon_0 = c^{-2}$.
[B] These quantities are used only in certain definitions. The local value of g should be used for precise experimental work.
[C] At 273.15 K and 101.325 kPa (STP).
[D] The exact value depends on the definition. The mean distance Earth–Sun given here is the astronomical unit.

[E] Maximum solar total radiant power above atmosphere at a distance of 1 AU from the Sun. Natural variability quoted.
[F] For London, 1960. Z positive downwards. Students are advised to ascertain the current local value.

References: CODATA Bulletin No. 11, Kaye, Weast.

The following provide a progression of lengths, masses, etc, which may be useful for estimations and checking calculations. Table 7.12 'Temperatures and the 1968 International Practical Temperature Scale' provides a progression of temperatures.

Orders of magnitude of physical quantities such as enthalpy or tensile strength may be obtained by consulting the appropriate tables.

LENGTH

10^{-15} m	radius of proton
10^{-12} m	wavelength of gamma ray (0.8 MeV)
10^{-10} m	lower limit of resolution of electron microscope
10^{-10} m	diameter of hydrogen atom
10^{-7} m	mean free path of air molecule (S.T.P.)
5×10^{-7} m	wavelength of visible light
10^{-6} m	diameter of finest drawn quartz fibre
2×10^{-6} m	diameter of staphylococcus (small bacterium)
5×10^{-6} m	length of human chromosome
7.5×10^{-6} m	diameter of human blood corpuscle
2×10^{-5} m	diameter of finest commercial glass capillary
10^{-4} m	thickness of paper (this book)
2×10^{-4} m	diameter of single strand of lighting flex
10^{-3} m	diameter of single stranded 5 A conductor
1.4×10^{-3} m	thickness of one penny piece
0.02 m	{ width of stamp / diameter of one penny piece
0.03 m	diameter of fifty penny piece
$0.212 \text{ m} \times 0.3$ m	A 4 paper[A]
0.25 m	length of standard brick
0.3 m	recommended 'module' for house building
1.8 m	height of man
5.0 m	height of modern two storey house to eaves
20 m	length of cricket pitch
294 m	length of R.M.S. *Queen Elizabeth II*
300 m	wavelength of radio waves (1 MHz)
450 m	height of Empire State Building
8800 m	height of Mount Everest
10^4 m	maximum depth of the ocean
1.7×10^6 m	radius of Moon
6×10^6 m	radius of Earth
4×10^8 m	distance Earth–Moon
1.5×10^{11} m	distance Earth–Sun
9.5×10^{15} m	1 light year
4.6×10^{16} m	distance of nearest star
10^{21} m	radius of local galaxy (Milky Way)
10^{22} m	average distance between galaxies
3×10^{26} m	radius of observable universe

SPEED

3×10^8 m s^{-1}	light in vacuum
6×10^5 m s^{-1}	electron (1 eV)
1.1×10^4 m s^{-1}	escape velocity from earth
2×10^3 m s^{-1}	neutron (0.025 eV)
500 m s^{-1}	air molecule (S.T.P.)
330 m s^{-1}	speed of sound in air (S.T.P.)
331 m s^{-1}	land speed record (740 mph, *Budweiser rocket*, 1979)
154 m s^{-1}	water speed record (345 mph, *Spirit of Australia*, 1977)
47 m s^{-1}	fastest bird (Peregrine falcon swooping)
33 m s^{-1}	speed limit on motorways (70 mph)
27 m s^{-1}	fastest land animal (Cheetah)
12 m s^{-1}	fastest sprint (Man)

A Successive paper sizes in the A range halve the longest dimension. A 4 is the metric size of paper which will be increasingly used for school work.

1

TIME

3×10^{-24}	s	light crossing proton
10^{-22}	s	proton revolution within nucleus
3×10^{-19}	s	light crossing atom
2×10^{-15}	s	period of visible light
10^{-13}	s	vibration period of ion in solid or atom in molecule
10^{-11}	s	period of millimetric wave
10^{-10}	s	period of molecular rotation
10^{-9}	s	lower limit of direct timing
10^{-8}	s	light crosses a room
10^{-7}	s	dead time of scintillation counter
10^{-6}	s	period of medium wave radio signal
10^{-5}	s	stroboscopic light flash
2×10^{-5}	s	dead time of geiger counter
10^{-4}	s	period of sound (highest audible frequency)
2×10^{-2}	s	period of a.c. mains oscillation
0.1 to 0.2	s	human reaction time
60	s	1 minute
500	s	light from Sun to Earth
10^5	s	1 day
3×10^7	s	1 year
2×10^9	s	human life span
5×10^{10}	s	half life of radium
10^{11}	s	oldest tree (bristlecone pine, California)
10^{14}	s	antiquity of man (*Homo sapiens*)
10^{17}	s	age of Earth and of oldest rocks
1.4×10^{17}	s	half life of uranium-238
10^{18}	s	expected life of Sun as a bright star

MASS

10^{-30}	kg	electron
1.7×10^{-27}	kg	proton
4×10^{-25}	kg	uranium atom
10^{-22}	kg	haemoglobin molecule
4×10^{-15}	kg	staphylococcus
6×10^{-10}	kg	limit of direct weighing
10^{-7}	kg	grain of sand (2×10^{-4} m radius)
2.5×10^{-3}	kg	smallest English mammal (pygmy shrew)
1.5×10^{-2}	kg	house mouse
4	kg	standard house brick
65	kg	man
100	kg	first earth satellite (Sputnik I, 1957)
10^3	kg	cubic metre of water
1.4×10^3	kg	car (*Maestro*)
10^4	kg	elephant
1.4×10^5	kg	{ heaviest object orbited, Apollo 15, 1971 / *Brachiosaurus* (Dry Mesa Quarry, Colorado)
1.9×10^5	kg	blue whale
3×10^6	kg	heaviest tree
3×10^8	kg	laden oil super-tanker
5×10^{18}	kg	total mass of atmosphere
10^{21}	kg	total mass of oceans
7×10^{22}	kg	mass of Moon
6×10^{24}	kg	mass of Earth
2×10^{30}	kg	mass of Sun
10^{41}	kg	mass of local galaxy (Milky Way)
10^{52}	kg	total mass of observable universe

The following information is given in the table.

1 Element. The elements are listed in alphabetical order. A name in italics means the element is not found naturally on Earth.

2 Z Atomic number.

3 A Molar mass for the naturally occurring isotopic composition to maximum precision so far attained.

To convert A to the old scale based on $A(^{16}O) = 16 \, g \, mol^{-1}$ exactly, multiply by 1.000 320 3. Apart from the elements listed in (i) and (ii) below, it is believed that the naturally occurring isotopic composition, and hence the value of A, is constant for each element throughout the Solar system and possibly throughout the Universe.

The following elements, marked * in the table, are found to exhibit variability in A due to variation in isotopic composition.

(i)
H	B	C	O	Si	S	
±0.000 01	±0.003	±0.000 05	±0.000 1	±0.001	±0.003	g mol^{-1}

The following elements, marked ‡ in the table, have rather large experimental uncertainties in A.

(ii)
Ne	Cl	Cr	Fe	Cu	Br	Ag	
±0.003	±0.001	±0.001	±0.003	±0.001	±0.002	±0.003	g mol^{-1}

4 Stable mass numbers and percentage abundances. Mass numbers of stable nuclides are given in bold type, with the percentage abundance in brackets. A mass number in italics means that the nuclide is radioactive. NR stands for naturally occurring radionuclide. *For radioactive elements only*, whether occurring in natural or artificial decay chains (see Table 2.3), AR stands for artificial radionuclide, usually that with longest half-life. There are also a large number of artificial radionuclides for the stable elements. Some of these (the more interesting or important) are given in Table 2.2 along with all nuclides up to $_8O$, and members of the ^{232}Th decay chain.

5 r Atomic radius in this table is the covalent radius where this is applicable, and otherwise the metallic radius. It gives an indication of atomic size. See Table 4.4 'Atomic radii and electronegativities' for fuller details of various atomic and ionic radii of elements.

6 $Q_{r\oplus}$ Terrestrial abundance of element by mass relative to that of silicon: that is, $Q_{r\oplus}(Si) = 100$. The absolute terrestrial abundance of silicon is believed to be 27.72 per cent by mass. The values given relate to the whole Earth and there are very significant local variations. Some of the figures quoted may be in error by a factor of 10 or more. See also Table 2.4.

In other parts of the Universe, the relative abundance of the less volatile elements probably approximates to that on Earth, but the relative abundances of gases and volatile elements depend very much on the size and thermal history of the body containing them. The following Solar abundances, also relative to silicon ($Q_{r\odot}(Si) = 100$), probably approximate to the values for the Universe as a whole.

Element	H	He	C	N	O	Mg	Fe
$Q_{r\odot}$	110 000	91 000	730	150	1700	68	24

7 p Price, approximate, in 1982–3. The price of a substance depends on its rarity, availability, cost of extraction, purity, physical state, and the amount supplied. The prices here are for small consignments (1 g to 1 kg) of cheapest 'chemical' purity (99% or better) from the 1982 catalogue of BDH Ltd. They price gases by volume: the bottling cost is about £10 per 40 dm^3. A figure in brackets after the main price is the commercial price (per kg) for large consignments (tonnes), taken from the FT index or from Prestel, January 1983, or from BOC Ltd 1982 catalogue. Radiochemical prices are from Amersham International PLC. The prices quoted for materials supplied as oxide or chloride are converted to price per kg of element.

References: Kaye, Lederer, and as Table 4.4 for r.

Element		Z	A/g mol⁻¹	Stable mass numbers and percentage abundances (in brackets)	r/nm	$Q_{r\oplus}$	p/£ kg⁻¹
Actinium	Ac	89	227.0278	227 (NR), 228 (NR), 225 (AR) (11 isotopes known)	—	1.3×10^{-15}	—
Aluminium	Al	13	26.9185	27(100)	0.130	35.8	12(0.7)
Americium	Am	95	243.0614	243 (AR) (10 isotopes known)	—	—	2.8/mCi ²⁴¹Am
Antimony	Sb[A]	51	121.75	121(57.25), 123(42.75)	0.143	4.4×10^{-4}	15(1.2†)
Argon	Ar[B]	18	39.948	36(0.34), 38(0.063), 40(99.6)	0.095	1.8×10^{-5}	116(0.3)
Astatine	At	85	209.9870	210 (AR), 217 (AR) (23 isotopes known)	—	—	16
Arsenic	As	33	74.9216	75(100)	0.122	2.2×10^{-3}	16
Barium	Ba	56	137.33	130(0.101), 132(0.097), 134(2.42), 135(6.59), 136(7.81), 137(11.32), 138(71.66)	0.198	5.7×10^{-1}	252
Berkelium	Bk	97	247.0703	247 (AR) (9 isotopes known)	—	—	—
Beryllium	Be	4	9.0122	9(100)	0.125	2.6×10^{-3}	1090[O]
Bismuth	Bi	83	208.9804	209(100)	0.152	8.8×10^{-5}	100(1.6)
Boron	B	5	10.81*	10(19.7*), 11(80.3*)	0.090	1.3×10^{-3}	22[O]
Bromine	Br	35	79.909‡	79(50.52), 81(49.48)	0.114	7.1×10^{-4}	5
Cadmium	Cd	48	112.41	106(1.22), 108(0.88), 110(12.39), 111(12.75), 112(24.07), 113(12.26), 114(28.86), 116(7.58)	0.141	6.6×10^{-5}	47(0.8)
Caesium[C]	Cs	55	132.9054	133(100)	0.235	3.1×10^{-3}	171[Cl]
Calcium	Ca	20	40.08	40(96.97), 42(0.64), 43(0.15), 44(2.06), 46(0.003), 48(0.19)	0.174	16.0	36
Californium	Cf	98	252.0817	252 (AR) (11 isotopes known)	—	—	—
Carbon	C	6	12.0111*	12(98.89), 13(1.11) {limestone CO₂}; 14(**NR)	0.077	1.4×10^{-1}	5[E]
Cerium	Ce	58	140.12	136(0.193), 138(0.23), 140(88.48), 142(11.07)	—	2.0×10^{-2}	464
Chlorine	Cl	17	35.453‡	35(75.53), 37(24.47)	0.099	1.4×10^{-1}	59
Chromium	Cr	24	51.996‡	50(4.31), 52(83.76), 53(9.55), 54(2.38)	0.117	4.4×10^{-2}	39
Cobalt	Co	27	58.9332	59(100)	0.126	1.0×10^{-2}	102(7)
Columbium	Cb now known as niobium						
Copper	Cu[D]	29	63.546‡	63(69.1), 65(30.9)	0.117	3.1×10^{-2}	10(1†)
Curium	Cm	96	247.0704	247 (AR) (13 isotopes known)	—	—	—
Deuterium	D	1	2.0141	Synonym for ²H	—	—	9080
Dysprosium	Dy	66	162.50	156(0.05), 158(0.09), 160(2.29), 161(18.88), 162(25.53), 163(24.97), 164(28.18)	—	8.5×10^{-5}	3110[O]
						2.0×10^{-3}	
Einsteinium	Es	99	254.088	254 (AR) (11 isotopes known)	—	—	—
Emanation	Em	obsolete for radon					
Erbium	Er	68	167.26	162(0.14), 164(1.56), 166(33.41), 167(22.94), 168(27.07), 170(14.88)	—	1.1×10^{-3}	2770[O]

* Variable isotopic composition, see introduction.
‡ Experimental uncertainty, see introduction. ᴬ Stibium.
ᴮ Formerly A was used to denote Argon.

ᶜ The spelling cesium is frequently used. ᴰ Cuprum.
† Price subject to violent fluctuations. ᴼ Supplied as oxide.
Cl Supplied as chloride. ᴱ Wood charcoal.

2

Element		Z	A/g mol⁻¹	Stable mass numbers and percentage abundances (in brackets)	r/nm	$Q_{r\oplus}$	ρ/£ kg⁻¹
Europium	Eu	63	151.96	151(47.77), 153(52.23)	—	4.7×10^{-4}	24 700[O]
Fermium	Fm	100	253.086	253 (AR) (10 isotopes known)	—	—	—
Fluorine	F	9	18.9984	19(100)	0.071	4.0×10^{-1}	5[H]
Francium	Fr	87	223.0198	223 (NR), 221 (AR), (18 isotopes known)	—	—	—
Gadolinium	Gd	64	157.25	152(0.20), 154(2.15), 155(14.7), 156(20.47), 157(15.68), 158(24.9), 160(21.9)	—	2.1×10^{-7}	2900[O]
Gallium	Ga	31	69.735	69(60.2), 71(39.8)	0.12	6.6×10^{-3}	7080
Germanium	Ge	32	72.59	70(20.55), 72(27.37), 73(7.67), 74(36.74), 76(7.67)	0.122	3.1×10^{-3}	4220
Gold	Au[D]	79	196.9665	197(100)	0.134	2.2×10^{-6}	29 600 (10 000[P])
Hafnium	Hf	72	178.49	174(0.16), 176(5.21), 177(18.56), 178(27.1), 179(13.75), 180(35.22)	0.144	2.0×10^{-3}	560[O]
Hahnium	Ha	105		AR (2 isotopes claimed)	—	—	—
Helium	He	2	4.0026	3(0.000 13) {atmosphere}, 4(≈100)	0.05	1.3×10^{-6}	1650 (53)
Holmium	Ho	67	164.9304	165(100)	—	5.1×10^{-4}	7045[O]
Hydrogen	H	1	1.0079*	1(99.985), 2(0.015), 3(**NR)	0.037	5.7×10^{-1}	3000 (2)
Indium	In	49	114.82	113(4.23), 115(95.77)	0.150	4.4×10^{-5}	977
Iodine	I	53	126.9045	127(100)	0.133	1.3×10^{-4}	18
Ionium	Io			obsolete for ²³⁰Th			
Iridium	Ir	77	192.22	191(38.5), 193(61.5)	0.126	4.4×10^{-7}	60 000
Iron	Fe[E]	26	55.847‡	54(5.84), 56(91.68), 57(2.17), 58(0.31)	0.125	22.0	1.5 (0.40)
Krypton	Kr	36	83.80	78(0.35), 80(2.27), 82(11.56), 83(11.55), 84(56.90), 86(17.37)	0.11	4.3×10^{-8}	545 (481)
Lanthanum	La	57	138.9055	138(0.09), 139(99.91)	0.169	8.1×10^{-3}	109[O]
Lawrencium	Lr	103	257	257 (AR) (3 isotopes known)	—	—	—
Lead	Pb[F]	82	207.19**	202(0.5), 204(1.40), 206(25.1), 207(21.7), 208(52.3)	0.154	7.0×10^{-3}	8.8 (0.3)
Lithium	Li	3	6.939	6(7.42), 7(92.58)	0.157	2.9×10^{-2}	207
Lutetium	Lu	71	174.97	175(97.4), 176(2.60)		3.3×10^{-4}	53 000[O]
Magnesium	Mg	12	24.312	24(78.60), 25(10.11), 26(11.29)	0.160	9.2	10 (1.4)
Manganese	Mn	25	54.9380	55(100)	0.139	4.4×10^{-1}	10 (0.7)
Mendelevium	Md	101	257.096	257 (AR) (3 isotopes known)	—	—	—
Mercury	Hg[G]	80	200.59	196(0.15), 198(10.02), 199(16.84), 200(23.13), 201(13.22), 202(29.80), 204(6.85)	0.144	2.2×10^{-4}	30 (30)
Molybdenum	Mo	42	95.94	92(15.86), 94(9.12), 95(15.70), 96(16.50), 97(9.45), 98(23.75), 100(9.62)	0.129	6.6×10^{-3}	120

* Variable isotopic composition, see introduction.
** Variable because of radioactivity.
‡ Experimental uncertainty, see introduction.

[D] Aurum. [E] Ferrum. [F] Plumbum. [G] Hydrargyrum. [H] As HF solution.
[O] Supplied as oxide. [P] Price fixed by financial policy.

Element		Z	A/g mol^{-1}	Stable mass numbers and percentage abundances (in brackets)	r/nm	$Q_{r\oplus}$	p/£ kg^{-1}
Neodymium	Nd	60	144.24	142(27.13), 143(12.20), 144(23.87), 145(8.29), 146(17.18), 148(5.72), 150(5.6)	—	1.1×10^{-2}	1124°
Neon	Ne	10	20.18‡	20(90.92), 21(0.26), 22(8.82)		3.1×10^{-8}	1820(720)
Neptunium	Np	93	239.0530^S	237 (AR), 239 (AR) (13 isotopes known)			95/mg ^{237}Np
Nickel	Ni	28	58.71	58(67.76), 60(26.16), 61(1.25), 62(3.66), 64(1.16)	0.121	3.5×10^{-2}	17(2.5)
Niobium	Nb	41	92.9064	93(100)	0.134	1.1×10^{-2}	615
Nitrogen	N	7	14.0067	14(99.63), 15(0.37)	0.075	9.0×10^{-2}	213(0.01)
Nobelium	No	102	255.093	255 (AR) (7 isotopes known)			
Osmium	Os	76	190.2	188(13.3), 189(16.1), 190(26.4), 192(41.0)	0.126	2.2×10^{-6}	17 600°
Oxygen	O	8	15.9994*	16(99.759), 17(0.037), 18(0.204) {air}	0.073	2.1×10^2	186(0.025)
Palladium	Pd	46	106.4	102(1.0), 104(11.0), 105(22.2), 106(27.3), 108(26.7), 110(11.8)	0.128	4.4×10^{-6}	17 700
Phosphorus	P	15	30.9738	31(100)	0.110	5.2	18^P
Platinum	Pt	78	195.09	190(0.01), 192(0.78), 194(32.9), 195(33.8), 196(25.2), 198(7.2)	0.129	2.2×10^{-6}	26 000 (6110†)
Plutonium	Pu	94	239.0522^T	238 (AR), 239 (AR), 242 (AR) (15 isotopes known)			540^R/μCi ^{239}Pu
Polonium	Po	84	210.0000	210 (NR) (27 isotopes known)		1.3×10^{-1}	
Potassium	K^H	19	39.0983	39(93.22), 40(0.12), 41(6.77)	0.235	11.4	52
Praseodymium	Pr	59	140.9077	141(100)		2.4×10^{-3}	1550
Promethium	Pm	61	144.9126	145 (AR), (14 isotopes known)			
Protactinium	Pa	91	231.0359	231 (NR), 233 (AR), 234 (NR) (14 isotopes known)		3.5×10^{-10}	
Radium	Ra	88	226.0254	226 (NR), 228 (NR), 224 (NR), 223 (NR) {order of decreasing abundance} (7 AR isotopes known)		5.7×10^{-9}	special order
Radon	Rn	86	222.0176	222 (NR), 220 (NR) {order of decreasing abundance} (22 isotopes known)	0.145	—	
Rhenium	Re	75	186.207	185(37.07), 187(62.93)	0.128	4.4×10^{-8}	11 600
Rhodium	Rh	45	102.9055	103(100)	0.125	4.4×10^{-7}	37 000
Rubidium	Rb	37	85.468	85(72.15), 87(27.85)	0.216	1.4×10^{-1}	1670^{Cl}
Ruthenium	Ru	44	101.07	96(5.46), 98(1.87), 99(12.63), 100(12.53), 101(17.02), 102(31.60), 104(18.87)	0.124	18×10^{-6}	15 300°
Rutherfordium	Rf	104	260	260 (AR) (3 isotopes claimed) Also called Kurchatovium			
Samarium	Sm	62	150.35	144(3.16), 147(15.07), 148(11.27), 149(13.82), 150(7.47), 152(26.63), 154(22.53)	—	2.8×10^{-3}	1620°

* Variable isotopic composition, see introduction.
† Price subject to violent fluctuations.
‡ Experimental uncertainty, see introduction.

H Kalium.　O Supplied as oxide.　P Supplied as red phosphorus.
R There is no free market in Pu.　S Precursor of ^{239}Pu.
T Principal fissile isotope, not most stable.　Cl Supplied as chloride.

2

Element		Z	$A/\text{g mol}^{-1}$	Stable mass numbers and percentage abundances (in brackets)	r/nm	$Q_{r\oplus}$	$p/\text{£ kg}^{-1}$
Scandium	Sc	21	44.9559	45(100)	0.144	2.2×10^{-3}	28 200°
Selenium	Se	34	78.96	74(0.89), 76(9.02), 77(7.58), 78(23.52), 80(49.82), 82(9.19)	0.117	4.0×10^{-5}	246(4.4)
Silicon	Si	14	28.0855*	28(92.18), 29(4.71), 30(3.12)	0.118	1.0×10^{2}	9(0.6)
Silver	Ag^{J}	47	107.868‡	107(51.35), 109(48.65)	0.134	4.4×10^{-5}	920(262†)
Sodium	Na^{K}	11	22.9898	23(100)	0.191	12.5	7
Strontium	Sr	38	87.62	84(0.56), 86(9.86), 87(7.02), 88(82.56)	0.191	1.3×10^{-1}	40^{Cl}
Sulphur	S	16	32.06*	32(95), 33(0.76), 34(4.22), 36(0.01)	0.102	2.3×10^{-1}	2.3
Tantalum	Ta	73	180.9479	180(0.01), 181(99.99)	0.134	9.2×10^{-4}	740
Technetium	Tc	43	98.9062	99 (AR) (16 isotopes known)	—		55/mCi ^{99}Tc
Tellurium	Te	52	127.60	120(0.09), 122(2.46), 123(0.87), 124(4.61), 125(6.99), 126(18.71), 128(31.79), 130(34.49)	0.135	8.8×10^{-7}	87
Terbium	Tb	65	158.9254	159(100)	—	4.0×10^{-4}	16 000°
Thallium	Tl	81	204.37	203(29.5), 205(70.5)	0.155	1.3×10^{-3}	77
Thorium	Th	90	232.038	232 (100 NR precursor) (Also decay products and 6 AR isotopes)		5.1×10^{-3}	117°
Thoron	(Tn)			synonym for ^{220}Rn			
Thulium	Tm	69	168.9342	169(100)	—	8.8×10^{-5}	48 100°
Tin	Sn^{L}	50	118.69	112(0.95), 114(0.65), 115(0.34), 116(14.24), 117(7.57), 118(24.01), 119(8.58), 120(32.97), 122(4.71), 124(5.98)	0.140	1.8×10^{-2}	26(7.7)
Titanium	Ti	22	47.90	46(7.99), 47(7.32), 48(73.99), 49(5.46), 50(5.25)	0.132	1.4	52
Tungsten	W^{M}	74	183.85	180(0.14), 182(26.4), 183(14.4), 184(30.6), 186(28.4)	0.130	3.0×10^{-2}	54(5.7)
Tritium	T	1	3.0161	3 (NR and AR) Synonym for ^{3}H.	—		35/Ci ^{3}H
Uranium	U	92	238.029	234(0.0057), 235(0.7196), 238(99.276) (Also 9 AR isotopes known) {all NR: proportions in natural U}	—	1.8×10^{-3}	44°
Vanadium	V	23	50.9415	50(0.25), 51(99.75)	0.122	6.6×10^{-2}	50°
Xenon	Xe	54	131.30	124(0.013), 126(0.09), 128(1.92), 129(26.44), 130(4.08), 131(21.18), 132(26.89), 134(10.4), 136(8.87)	0.130	5.2×10^{-10}	1280(870)
Ytterbium	Yb	70	173.04	168(0.14), 170(3.03), 171(14.31), 172(21.82), 173(16.13), 174(31.84), 176(12.73)	—	1.2×10^{-3}	6640°
Yttrium	Y	39	88.9059	89(100)	0.162	1.2×10^{-2}	1472°
Zinc	Zn	30	65.38	64(48.89), 66(27.81), 67(4.11), 68(18.56), 70(0.62)	0.12	5.8×10^{-2}	3.9(0.5)
Zirconium	Zr	40	91.22	90(51.46), 91(11.23), 92(17.11), 94(17.40), 96(2.80)	0.145	9.7×10^{-2}	16

* Variable isotopic composition, see introduction.
† Price subject to violent fluctuations.
‡ Experimental uncertainty, see introduction.

Cl Supplied as chloride. J Argentum.
K Natrium. L Stannum.
M Wolfram. O Supplied as oxide.

This table contains all known isotopes of the elements up to $_8$O (note there is no nuclide of mass 5), all members of the ^{232}Th decay chain (see Table 2.3), some fission products, and other interesting or important nuclides. For details of mass numbers and percentage abundances of naturally occurring isotopes of the elements see Table 2.1.

There are about 1600 different nuclides (some doubtful). 238 are stable isotopes of 80 elements, 49 are very slightly radioactive isotopes of elements up to ^{209}Bi (30 have half-lives in excess of 10^{13} years), and 39 occur in naturally occurring radioactive decay chains (see Table 2.3). The remainder have been produced artificially by various forms of nuclear reaction. 102 are isotopes of transuranic elements and 41 are isotopes of Tc and Pm; none of these elements occurs naturally. There are also over 200 isomeric (metastable) states of nuclides with half-lives in excess of 1 second. ^{3}H (^{3}T) and ^{14}C are produced in the atmosphere by cosmic ray bombardment.

The following information is given in the table. (Doubtful figures are in italics.)

1 Z **Atomic number**, that is, the number of protons in the nucleus and the number of electrons in the outer part of the atom.

2 A **Mass or nucleon number**, that is, the number of protons and neutrons in the nucleus. (Hence $N = A - Z$ gives the number of neutrons.) C indicates commercial availability.

3 M **Exact mass of the nuclide including electrons**, measured in atomic mass units (u), where $u = 1.66 \times 10^{-27}$ kg.

4 Δ **Energy equivalent of the mass defect or excess** on a scale where Δ for ^{12}C is zero. $\Delta = (M - A)c^2$. Changes in Δ can be used to calculate the energy released in nuclear reactions, or vice versa. Many of the exact masses given here were obtained from measured values of Δ. The lower the value of Δ, the more stable the nuclide.

5 B **Total binding energy of the nucleus**. $B = (M - Zm_p - (A - Z)m_n)c^2$.

6 a **Abundance of stable or nearly stable nuclides** as a percentage of the total number of atoms of that element on Earth.

7 $T_{\frac{1}{2}}$ **Half-life of radioactive nuclide.**

8 Decay. Only the principal modes of decay are given, with E/MeV in parentheses, where E stands for the maximum particle energy released in that mode or the most intense γ energy. Where appropriate the percentage of decays via in this mode is given. The following abbreviations are used.

α   Emission of alpha particle. $\beta^{\pm}$ Emission of positive or negative beta particle.
Ec Electron capture (generally an alternative to β^+ emission, always results in γ emission).
e^- Ejection of atomic electron by nuclear γ-ray (internal conversion).
γ   Emission of γ-ray photon.   SF Spontaneous fission.
IT Isomeric transition from an excited state to a lower state of the same nuclide by γ emission.

The following symbolism is used to show the relationship of one decay mode to another.
, The first decay sometimes leaves the nucleus in an excited state which decays by the next mode.
; The first emission is invariably followed by the second.   : The two modes are alternative decay paths.

The decay of many nuclides is quite complex and may involve several different γ-rays or even several successive γ-rays, with consequent variations in the particle energy. It should also be noted that β emission is accompanied by a (virtually undetectable) neutrino, so that β-particle energy is variable with a continuous spectrum. Values quoted are the maximum possible.

9 Production. Only one method by which the nuclide has been produced has been given. The notation is:
parent nuclide (bombarding particle, outgoing particle)
The following abbreviations are used: d   deuteron, that is, ^{2}H;   t   triton, that is ^{3}H;   NR   natural radioactive nuclide;   D daughter product of following nuclide;   fiss occurs as the result of ^{235}U fission.

10 σ **Absorption cross-section for thermal neutrons**, that is, a measure of how readily neutrons are absorbed by the nuclide. In general, absorption of a neutron results in the emission of a γ-ray and the formation of a heavier (and possibly radioactive) isotope of the same element – the (n, γ) reaction. It may, however, cause the emission of one or more charged particles, or fission. σ gives the total absorption cross-section unless a specific reaction is noted.

Z	A	M/u	Δ/MeV	B/MeV	a/%	$T_{\frac{1}{2}}$	Decay	Production	$\sigma/10^{-28} \text{ m}^2$
-1 e	0	0.000549	0.510^A	—	—	—	stable	—	—
0 n	1	1.008665	8.071	-0.552	—	11.7 min	β^-(0.78)	^{9}Be(α, n)	0
1 p	1	1.007276	6.777	—	—	—	stable	—	0.332
1 H	1^s	1.007825	7.289	—	99.985*	—	stable	—	0.332
	2	2.014102	13.136	-1.713	0.015*	—	stable	—	0.0005
	3 C	3.016050	14.950	-7.970	—	12.3 a	β^-(0.0186) no γ	^{6}Li(n, α)	$<7 \times 10^{-6}$
2 He	3	3.016030	14.931	-6.695	0.00013	—	stable	^{3}H(—, β^-)	5330
	4^s	4.002603	2.425	-27.273	99.99987	—	stable	—	0
	6	6.018893	17.598	-28.241	—	0.8 s	β^-(3.51)	^{7}Li(γ, p)	—
	8	8.034	31.7	-30.3	—	0.12 s	β^-(9.7); γ(0.98) 98%: n	^{12}C(p, ?)	—
3 Li	6^s	6.015124	14.088	-30.458	7.42	—	stable	—	950
	7	7.016004	14.907	-37.710	92.58	—	stable	—	0.037
	8	8.022487	20.946	-39.742	—	0.85 s	β^-(13); 2α(1.6)	^{7}Li(n, γ)	—
	9	9.02680	24.97	-43.796	—	0.17 s	β^-(13.6) 25%: β^-(11.2); n; 2α	^{9}Be(n, p)	—
4 Be	6	6.01972	18.37	-24.883	—	0.4 s	'particle unstable'	^{9}Be(p, ?)	—
	7 C	7.016929	15.769	-35.555	—	53.6 d	Ec, γ(0.477) 10%	^{10}B(p, α)	51000
	9^s	9.012186	11.351	-56.115	100	—	stable	—	0.009
	10	10.013534	12.607	-62.931	—	2.7×10^6 a	β^-(0.555) no γ	^{9}Be(d, p)	<0.001
	11	11.02167	20.18	-63.423	—	13.6 s	β^-(11.5) 61%, γ(2.12) 32%	^{11}B(n, p)	—
	12	12.027	25	-66.530	—	0.0114 s	β^-(12), γ: n	^{18}O(p, ?)	—
5 B	8	8.024609	22.923	-35.178	—	0.77 s	β^+(14.0); 2α(1.6)	^{6}Li(^{3}He, n)	—
	10	10.012938	12.052	-62.192	19.78	—	stable	—	3837(n, α)
	11^s	11.009305	8.667	-73.647	80.22	—	stable	^{11}B(d, p)	0.005
	12	12.014354	13.370	-77.015	—	0.020 s	β^-(13.37) 98%, γ(4.43) 1.3%: α(0.195)		—
	13	13.01778	16.56	-81.895	—	0.0186 s	β^-(13.44) 93%, γ(3.68) 7%: n	^{11}B(t, p)	—

* Natural variation (<1%).

^{s} Most stable isotope of element on criterion of least Δ.

^{A} Energy equivalent of total mass of particle.

Z	A		M/u	Δ/MeV	B/MeV	a/%	$T_{\frac{1}{2}}$	Decay	Production	$\sigma/10^{-28}\ m^2$
6 C	9		9.03104	29.00	−35.965	—	0.127 s	β^+; p(8.2) 60%: p(1.1) 40%: 2α(0.05)	^{10}B(p, 2n)	—
	10		10.01686	15.70	−57.237	—	19.4 s	β^-(1.87); γ(0.72)	^{10}B(p, n)	—
	11		11.011432	10.648	−70.372	—	20.4 min	β^+(0.97) no γ	^{10}B(d, n)	—
	12[s]		12 (Def)	0 (Def)	−89.092	98.89*	stable			0.0034
	13		13.003354	3.124	−94.039	1.11*	stable			0.0009
	14	C	14.003242	3.020	−102.215	—	5730 a	β^-(0.156) no γ	^{14}N(n, p)	—
	15		15.010599	9.873	−103.433	—	2.4 s	β^-(9.82) 32%: β^-(4.52); γ(5.3) 68%	^{14}C(d, p)	—
	16		16.01470	13.69	−107.684	—	0.74 s	β^-(8.0); n	^{14}C(t, p)	—
7 N	12		12.01864	17.36	−70.435	—	0.011 s	β^+(16.4), 3α(0.195) 3%; γ(4.4) 3%	^{12}C(p, n)	—
	13		13.005738	5.345	−90.525	99.63	10.0 min	β^+(1.2) no γ	^{10}B(α, n)	1.81
	14		14.003074	2.864	−101.077	0.37	stable			2.4×10^{-5}
	15[s]		15.000108	0.101	−111.911	—	stable			—
	16		16.006103	5.685	−114.398	—	7.14 s	β^-(10.4), γ(6.13) 68%, α 0.001%	^{15}N(n, γ)	—
8 O	17		17.00845	7.87	−120.283	—	4.16 s	β^-(8.68): β^-(4.1); n 95%	^{14}C(α, p)	—
	18		18.0142	13.1	−122.999	—	0.63 s	β^-(9.4); γ(1.98)	^{18}O(n, p)	—
	13		13.0248	23.2	−71.475	—	0.0087 s	β^+; p(6.97) no γ	^{14}N(p, 2n)	—
	14		14.008597	8.008	−94.639	—	71 s	β^+(4.12), γ(2.31) 99%	^{14}N(p, n)	—
	15		15.003070	2.860	−107.859	—	2.06 min	β^+(1.74) no γ	^{14}N(d, n)	—
	16[s]		15.994915	−4.737	−123.526	99.759*	stable			0.0002
	17		16.999133	−0.807	−127.668	0.037*	stable			0.24 (n, α)
	18		17.999161	−0.782	−135.713	0.204*	stable			0.0002
	19		19.003578	3.333	−139.670	—	27 s	β^-(4.6); γ(0.197) 97%; γ	^{18}O(n, γ)	—
	20		20.00408	3.80	−147.274	—	14 s	β^-(2.75); γ(1.06)	^{18}O(t, p)	—
11 Na	22	C	21.994437	−5.182	−168.5	—	2.602 a	β^+(1.82), γ(1.27) 99.95%	^{19}F(α, n)	40000
	23[s]		22.989771	−9.528	−180.9	100	stable			0.53
	24	C	23.990964	−8.417	−187.9	—	14.96 h	β^-(1.39); γ(2.75); γ(1.37) 99%	^{23}Na(n, γ)	—
13 Al	27		26.981541	−17.194	−218.3	100	stable			0.232
15 P	30		29.97832	−20.19	−242.9	—	2.5 min	β^+(3.24), γ(2.23) 0.5%	^{27}Al(α, n)	—
	31[m]		30.973765	−24.438	−255.2	100	stable			0.19
	32	C	31.973909	−24.303	−263.2	—	14.3 d	β^-(1.71) no γ	^{31}P(n, γ)	—

* Natural variation (<1%).
s Most stable isotope of element on criterion of least Δ.
m Metastable isomer.

Z	A		M/u	Δ/MeV	B/MeV	$a/\%$	$T_{\frac{1}{2}}$	Decay	Production	$\sigma/10^{-28}\,\text{m}^2$
16 **S**	35	C	34.969033	−28.845	−290.6	—	87 d	$\beta^-(0.17)$ no γ	$^{34}\text{S}(n,\gamma)$	—
17 **Cl**	36	C	35.968307	−29.521	−298.1	—	3×10^5 a	$\beta^-(0.71)$: Ec(1.14) 1.9%: β^+ no γ	$^{35}\text{Cl}(n,\gamma)$	100
19 **K**	40		39.964000	−33.533	−331.8	0.118	1.3×10^9 a	$\beta^-(1.31)$: Ec(1.51) 11%; $\gamma(1.46)$: β^+	NR	70
26 **Fe**	54	C	53.93962	−56.24	−458.5	5.82	—	stable	—	2.8
	55	C	54.938295	−57.477	−467.0	—	2.60 a	Ec(0.23)	$^{54}\text{Fe}(n,\gamma)$	—
	56		55.934934	−60.608	−479.0	91.66	—	stable	—	2.6
	57	C	56.935391	−60.182	−486.6	2.19	—	stable	A	2.5
	58ˢ		57.933275	−62.153	−496.6	0.33	—	stable	—	1.2
	59	C	58.934879	−60.668	−503.2	—	45 d	$\beta^-(1.57)$ 0.3%; $\gamma(1.1)$ 56%	$^{58}\text{Fe}(n,\gamma)$	—
27 **Co**	56	C	55.93985	−56.03	−473.1	—	77 d	$\beta^+(1.49)$: Ec 80%; γ; $\gamma(0.84)$	$^{56}\text{Fe}(p,n)$	—
	57	C	56.936299	−59.346	−484.5	—	270 d	Ec(0.84); $\gamma(0.14)$, e^-	$^{55}\text{Mn}(\alpha, 2n)^\text{B}$	18 (37)†
	59		58.933189	−62.233	−503.5	100	—	stable	—	6 (2)†
	60	C	59.933811	−61.653	−511.0	—	5.26 a	$\beta^-(0.31)$; $\gamma(1.33)$; $\gamma(1.17)$	$^{59}\text{Co}(n,\gamma)$	—
28 **Ni**	64		63.92796	−67.10	−547.4	1.08	—	stable	—	1.5
29 **Cu**	64	C	63.929757	−65.430	−544.5	—	12.8 h	$\beta^-(0.57)$ 38%: $\beta^+(0.66)$ 19%: Ec	$^{63}\text{Cu}(n,\gamma)$	<6000
30 **Zn**	64ˢ	C	63.929140	−66.005	−543.7	48.89	—	stable	—	0.46 (0.8)†
	65	C	64.92923	−65.92	−551.7	—	244 d	$\beta^+(0.327)$, Ec 98%, e^- (1.11)	$^{64}\text{Zn}(n,\gamma)$	—
36 **Kr**	84ᵐ		83.911505	−82.431	−713.8	56.90	—	stable	—	0.13
	90		89.9197	−74.8	−754.6	—	33 s	$\beta^-(2.8)$, $\gamma(1.1)$ 48%; $\gamma(0.54)$	fiss	—
	91		90.923	−72	−759.6	—	9 s	$\beta^-(3.6)$, $\gamma(0.1)$	fiss	—
37 **Rb**	90		89.9148	−79.4	−757.9	—	2.6 min	$\beta^-(6.6)$, $\gamma(0.83)$	D ^{90}Kr	—
38 **Sr**	88ᵐ		87.905628	−87.906	−749.0	82.56	—	stable	—	0.005
	90		89.90775	−85.93	−763.2	—	28.1 a	$\beta^-(0.546)$ no γ	D ^{90}Rb	0.9
	93		92.9142	−79.9	−781.4	—	8 min	$\beta^-(3.9)$ 14%; $\gamma(0.6)$	fiss	—
	94		93.9154	−78.8	−788.3	—	1.4 min	$\beta^-(2.1)$; $\gamma(1.42)$	fiss	—

ᵐ Metastable isomer.
ˢ Most stable isotope of element on criterion of least Δ.
† Disagreement between sources.

A Mössbauer resonance γ absorption (from ^{57}Co).
B Mössbauer source.

Z	A	M/u	Δ/MeV	B/MeV	a/%	T½	Decay		
47 Ag	107	106.905091	−88.406	−891.2	51.82		stable		35
	108	107.905953	−87.603	−898.5		2.4 min	β⁻(1.64) 96%: Ec: β⁺(0.9), γ	D ¹⁰⁸ᵐAg	—
	108ᵐ	107.90606	−87.50	>−898.4		>5 a	Ec 90%; γ(0.72); 2γ: IT	¹⁰⁷Ag(n, γ)	—
	109ˢ	108.904756	−88.718	−907.7	48.18		stable		93 (3 to ¹¹⁰ᵐAg)
	110	109.906114	−87.453	−914.48		24.4 s	β⁻(2.87), γ(0.66) 4.5%	¹⁰⁹Ag(n, γ)	82
	110ᵐ	109.90622	−87.35	−914.38		253 d	β⁻(1.5), γ(0.89) 71%; γ(0.66) 96%: IT	¹⁰⁹Ag(n, γ)	
48 Cd	113	112.904408	−89.042	−939.0	12.26	3×10¹⁵ a	not known	(NR)	20000
49 In	115ˢ	114.90387	−89.54	−954.3	95.72	6×10¹⁴ a	β⁻(0.48) no γ	(NR)	45 (198)†
	116	115.90553	−88.25	−961.14		14 s	β⁻(3.3)	¹¹⁵In(n, γ)	—
	116ᵐ	115.90538	−88.14	−961.00		54 min	β⁻(1.0); γ; γ(1.3) 80%	c	—
53 I	127ᵐ	126.904474	−88.981	−1045.5	100		stable		6.2
	131 C	130.906127	−87.441	−1076.2		8.07 d	β⁻(0.61); γ(0.36) 82%	fiss	≈0.7
	135	134.91006	−83.77	−1104.8		6.7 h	β⁻(1.4); γ	fiss	—
54 Xe	130ᵐ	129.903509	−89.880	−1069.3	4.08		stable		18
	135	134.91350	−86.50	−1100.3		9.2 h	β⁻(0.91); γ(0.25), e⁻(0.2)	fiss, D ¹³⁵I	2.7×10⁶
	139	138.9184	−76.0	−1128.0		40 s	β⁻(4.6) 31%; γ	fiss	—
	140	139.921	−74	−1133.7		14 s	β⁻(4.7), γ; γ(0.22) 77%	fiss	—
55 Cs	130	129.90676	−86.85	−1065.0		29 min	Ec 53%: β⁺(1.97): β⁻(0.44) 2%	¹²⁷I(α, 2n)	29
	133(C)	132.90544	−88.08	−1090.4	100		stable	D	8.7
	135	134.90590	−87.65	−1106.1		3×10⁶ a	β⁻(0.21) no γ	fiss, D ¹³⁵Xe	8.7
	137 C	136.90707	−86.56	−1121.2		30.0 a	β⁻(1.18), γ(0.66) 85%	fiss	0.11
56 Ba	136ᵐ	135.90456	−88.90	−1114.1	7.81	12 s	β⁻(?4)		0.4
	143	142.921	−74	−1155.3			β⁻(?3)	fiss	—
	144	143.923	−72	−1161.5		11.4 s		fiss	—
81 Tl	208	207.98201	−16.76	−1590.8		3.1 min	β⁻(1.80); γ; γ(2.61)	D ²³²Th	—
82 Pb	206ˢ	205.97447	−23.78	−1580.4	23.6		stable	D ²³⁸U	0.03 (0.3)†
	207	206.975903	−22.446	−1587.1	22.6		stable	D ²³⁵U	0.71
	208	207.97666	−21.74	−1594.5	52.3		stable	D ²³²Th	0.0005 (0.015)†
	210 C	209.98420	−14.72	−1603.6		22 a	β⁻(0.06), γ(0.05) 81%: α(3.7) 10⁻⁶%	D ²³⁸U	—
	212	211.9189	−7.55	−1680.6		10.6 h	β⁻(0.58), γ(0.24) 81%	D ²³²Th	—
	214	213.99984	−0.148	−1621.3		26.8 min	β⁻(1.03), γ(0.35) 47%	—	—

ᵐ Metastable isomer.
ˢ Most stable isotope of element on criterion of least A, nevertheless can be radioactive.
† Disagreement between sources.

C Two isomeric states exist.
D Mössbauer resonance γ absorption (from ¹³³Ba).

Z	A		M/u	Δ/MeV	B/MeV	a/%	T½	Decay	Production	σ/10⁻²⁸ m²
83 Bi	212		211.99128	-8.12	-1611.8	—	60.6 min	β⁻(2.25), γ(0.7): α(6.09) 36%	D ²³²Th	—
84 Po	212		211.988865	-10.372	-1612.8	—	3 × 10⁻⁷ s	α(8.78)	D ²³²Th	—
	216		216.00192	1.79	-1632.9	—	0.145 s	α(6.78)	D ²³²Th	—
86 Rn	220		220.01139	10.62	-1653.8	—	55.5 s	α(6.29)	D ²³²Th	<0.2
	222		222.01761	16.40	-1664.1	—	3.82 d	α(5.49)	D ²³⁸U	0.7
88 Ra	224		224.02020	18.82	-1675.3	—	3.64 d	α(5.68), γ(0.24) 5.5%	D ²³²Th	12
	226		226.02544	23.69	-1686.5	—	1622 a	α(4.78), γ(0.19) 5.4%	D ²³⁸U	20
	228		228.03110	28.97	-1697.4	—	5.75 a	β⁻(0.05), γ(0.026) ? 30%	D ²³²Th	36
89 Ac	228		228.03104	28.91	-1696.2	—	6.13 h	β⁻(2.11); γ(0.06), γ(0.9)	D ²³²Th	—
90 Th	228	C	228.02873	26.76	-1697.0	—	1.91 a	α(5.43), γ(0.08) 28%	D ²³²Th	123
	230	C	230.03316	30.89	-1709.1	—	80000 a	α(4.68): α(4.62) 24%; γ(0.06)	D ²³⁸U	—
	232	C	232.03808	35.47	-1720.6	100	1.4 × 10¹⁰ a	α(4.01): α(3.95) 23%; γ(0.06)	NR	7.4
	234		234.04364	40.65	-1731.6	—	24.1 d	β⁻(0.19), γ(0.09)	D ²³⁸U	1.8
91 Pa	234ᵐ		234.04342	40.45	-1730.5	—	1.17 min	β⁻(2.29): IT	D ²³⁸U	—
92 U	233	C	233.03965	36.93	-1724.6	—	1.6 × 10⁵ a	α(4.82), γ(0.04) 15%	²³²Th(n, γ 2β⁻)	579, 532ᶠ
	234		234.04098	38.153	-1731.5	0.00572	2.47 × 10⁵ a	α(4.77), γ(0.05) 28%	D ²³⁴Pa	95
	235	C	235.04394	40.928	-1736.8	0.72	7.13 × 10⁸ a	α(4.6) 4.6%, γ(0.2)	NR	681, 582ᶠ
	236		236.04559	42.456	-1743.3	—	2.39 × 10⁷ a	α(4.49), γ(0.05) 26%	²³⁵U(n, γ)	6
	238		238.05082	47.327	-1754.6	99.27	4.51 × 10⁹ a	α(4.20), γ(0.05) 23%	NR	2.72
	239		239.05433	50.597	-1759.4	—	23.5 min	β⁻(1.29), γ(0.07) 74%	²³⁸U(n, γ)	36, 14ᶠ
93 Np	239		239.05295	49.311	-1759.4	—	2.35 d	β⁻(0.71); γ	D ²³⁹U	63
94 Pu	239		239.05218	48.594	-1758.8	—	24360 a	α(5.16); γ(0.00008), γ	D ²³⁹Np	1005, 763ᶠ
	240		240.05384	50.131	-1765.3	—	6580 a	α(5.17), γ(0.05) 24%	²³⁹Pu(n, γ)	290, 0.05ᶠ
	241		241.05687	52.972	-1770.6	—	13.27 a	β⁻(0.02): α(4.9) 0.002%	²⁴⁰Pu(n, γ)	1371, 1071ᶠ
95 Am	239		239.05304	49.405	-1756.7	—	12 h	Ec; γ(0.29), γ(0.23)	²³⁹Pu(p, n)	787, 3.3ᶠ
	241	C	241.05685	52.953	-1769.3	—	433 a	α(5.49), γ(0.06) 36%	D ²⁴¹Pu	—
100 Fm	249		249.079	73.790	-1806.8	—	2.5 min	α(7.9)	²³⁸U(¹⁶O, 5n)	—
	251		251.082	76.38	-1820.1	—	7 h	Ec: α(6.9)	²⁴⁹Cf(α, 3n)	—
104 Rfᴬ	260		—	—	—	—	0.3 s	SF?: α(9) ?	²⁴²Pu(²²Ne, 4n)	—

References: American Institute of Physics Handbook, Lederer.

† Disagreement between sources.

ᵐ Metastable isomers.   ᶠ Fission cross-section.
ᴬ Also known at Ku, kurchatovium.

The following diagrams show the processes which occur in various decay chains. Each diagram gives (a) the radiation emitted, (b) the half-life of the process, and (c) the average particle energy of the radiation. Branches involving less than 1% of the atoms have been omitted. Gamma radiation is emitted after the majority of the decays.

A mass number Z atomic number n integer

Thorium series $A = 4n$

Symbols in brackets are the old symbols for the nuclide (no longer used).

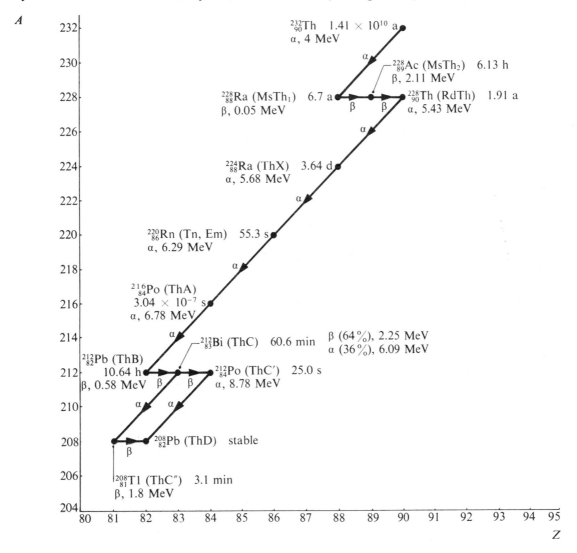

$^{232}_{90}$Th 1.41×10^{10} a
α, 4 MeV

$^{228}_{89}$Ac (MsTh₂)   6.13 h
β, 2.11 MeV

$^{228}_{88}$Ra (MsTh₁)   6.7 a
β, 0.05 MeV

$^{228}_{90}$Th (RdTh)   1.91 a
α, 5.43 MeV

$^{224}_{88}$Ra (ThX)   3.64 d
α, 5.68 MeV

$^{220}_{86}$Rn (Tn, Em)   55.3 s
α, 6.29 MeV

$^{216}_{84}$Po (ThA)
3.04×10^{-7} s
α, 6.78 MeV

$^{212}_{83}$Bi (ThC)   60.6 min

β (64%), 2.25 MeV
α (36%), 6.09 MeV

$^{212}_{82}$Pb (ThB)
10.64 h
β, 0.58 MeV

$^{212}_{84}$Po (ThC′)   25.0 s
α, 8.78 MeV

$^{208}_{82}$Pb (ThD)   stable

$^{208}_{81}$Tl (ThC″)   3.1 min
β, 1.8 MeV

Neptunium series $A = 4n + 1$

This series does not occur in nature. The precursors are formed in nuclear reactors.

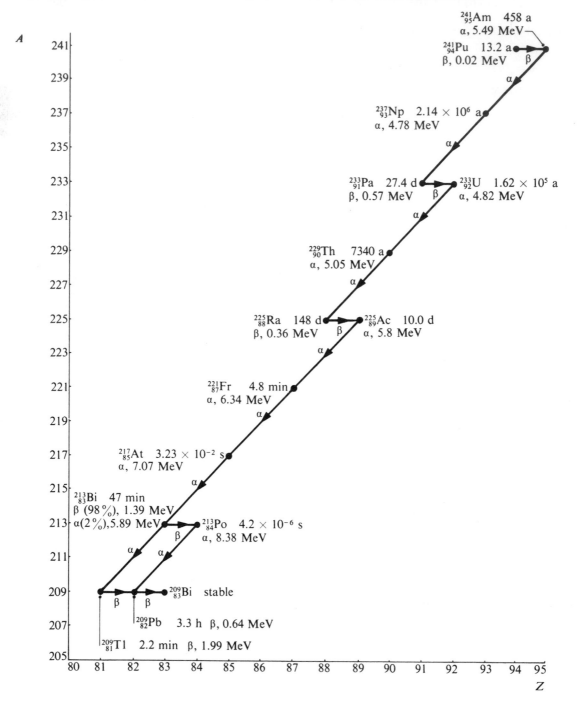

A

241 — $^{241}_{95}$Am   458 a
α, 5.49 MeV
$^{241}_{94}$Pu   13.2 a
β, 0.02 MeV β

237 — $^{237}_{93}$Np   2.14 × 10⁶ a
α, 4.78 MeV

233 — $^{233}_{91}$Pa   27.4 d $^{233}_{92}$U   1.62 × 10⁵ a
β, 0.57 MeV β α, 4.82 MeV

229 — $^{229}_{90}$Th   7340 a
α, 5.05 MeV

225 — $^{225}_{88}$Ra   148 d $^{225}_{89}$Ac   10.0 d
β, 0.36 MeV β α, 5.8 MeV

221 — $^{221}_{87}$Fr   4.8 min
α, 6.34 MeV

217 — $^{217}_{85}$At   3.23 × 10⁻² s
α, 7.07 MeV

213 — $^{213}_{83}$Bi   47 min
β (98 %), 1.39 MeV
α(2 %),5.89 MeV $^{213}_{84}$Po   4.2 × 10⁻⁶ s
α, 8.38 MeV

209 — $^{209}_{83}$Bi   stable
β β
$^{209}_{82}$Pb   3.3 h β, 0.64 MeV
$^{209}_{81}$Tl   2.2 min β, 1.99 MeV

80 81 82 83 84 85 86 87 88 89 90 91 92 93 94 95

Z

2

Uranium series $A = 4n + 2$

Symbols in brackets are the old symbols for the nuclide (no longer used).

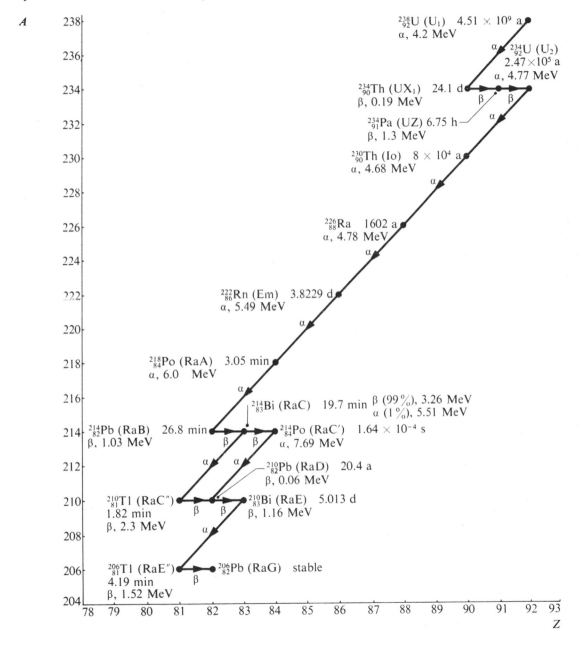

A

238 — $^{238}_{92}U$ (U₁) 4.51×10^9 a   α, 4.2 MeV

236 — α, $^{234}_{92}U$ (U₂) 2.47×10^5 a   α, 4.77 MeV

234 — $^{234}_{90}Th$ (UX₁)   24.1 d   β, 0.19 MeV   β   β

232 — $^{234}_{91}Pa$ (UZ) 6.75 h   β, 1.3 MeV   α

230 — $^{230}_{90}Th$ (Io) 8×10^4 a   α, 4.68 MeV   α

228

226 — $^{226}_{88}Ra$ 1602 a   α, 4.78 MeV   α

224

222 — $^{222}_{86}Rn$ (Em)   3.8229 d   α, 5.49 MeV   α

220

218 — $^{218}_{84}Po$ (RaA)   3.05 min   α, 6.0 MeV   α

216 — $^{214}_{83}Bi$ (RaC)   19.7 min   β (99%), 3.26 MeV   α (1%), 5.51 MeV

214 — $^{214}_{82}Pb$ (RaB)   26.8 min   β, 1.03 MeV   β   β $^{214}_{84}Po$ (RaC′) 1.64×10^{-4} s   α, 7.69 MeV

212 — α   α $^{210}_{82}Pb$ (RaD)   20.4 a   β, 0.06 MeV

210 — $^{210}_{81}Tl$ (RaC″)   1.82 min   β, 2.3 MeV   β   β $^{210}_{83}Bi$ (RaE)   5.013 d   β, 1.16 MeV

208 — α

206 — $^{206}_{81}Tl$ (RaE″)   4.19 min   β, 1.52 MeV   β $^{206}_{82}Pb$ (RaG)   stable

204

78   79   80   81   82   83   84   85   86   87   88   89   90   91   92   93

Z

Actinium series $A = 4n + 3$

Symbols in brackets are the old symbols for the nuclide (no longer used).

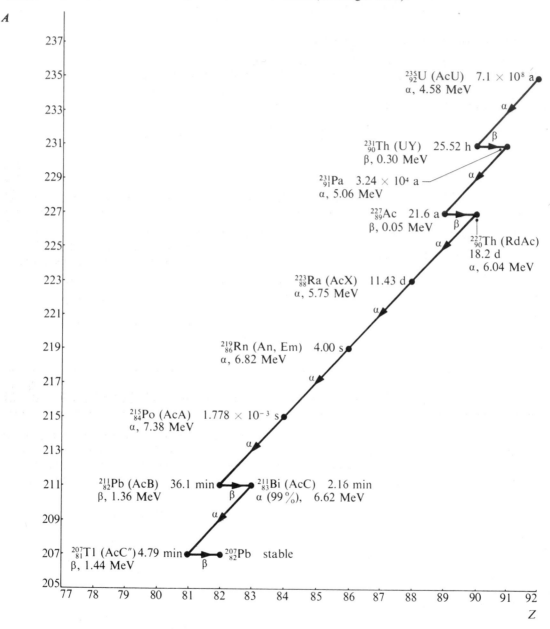

x Mass of given element in the Earth.

Element	O	Si	Al	Fe	Ca	Na	K	Mg	Ti	P	Mn	All other elements together
$x/\%$	46.6	27.7	8.1	5.0	3.6	2.8	2.6	2.1	0.4	0.1	0.1	<1

V_p Partial volume of constituent gas in Earth's atmosphere.

Gas	N_2	O_2	Ar	CO_2	Ne	He	CH_4	Kr	Xe
$V_p/\%$	78.09	20.95	0.93	0.03	0.0018	0.0005	0.0002	0.0001	0.00001

x Mass fraction of element in sea water.

Element	O	H	Cl	Na	Mg	S	Ca	K	Br	C
$x/\%$	85.7	10.8	1.90	1.1	0.14	0.09	0.04	0.04	0.007	0.003

2

RANGE OF α-PARTICLES IN AIR, AND ELECTRONS IN ALUMINIUM | **2·5**

E Initial energy of α-particle R Range in air

Nuclide	E/MeV	$R(\text{air, 288 K, 1 atm})/\text{cm}$
^{232}Th	4.00	2.59
^{238}U	4.20	2.67
^{226}Ra	4.77	3.39
^{210}Po	5.30	3.84
^{214}Po	7.68	6.95
^{212}Po	8.78	8.57

E Initial energy of electron
R Range in aluminium (expressed as length × mass concentration of Al in a non-absorber)

E/MeV	0.01	0.05	0.1	0.4	1.0
$R/\text{g cm}^{-2}$	0.16	4.0	13.5	120	420

E/MeV	2.0	5.0	10.0
$R/\text{g cm}^{-2}$	950	2540	5200

BIOLOGICALLY SIGNIFICANT LEVELS OF IONIZING RADIATION | **2·6**

1 mSv (100 mRem) per individual averaged over a whole population is believed to result in an average total of 13 cancers and 8 genetic defects per million of the population in later years. (The spontaneous incidence of cancer is about 200 per million per year.) Whole life exposure reduces life span by about one year per sievert. Permissible dose rates are based on these estimates, which are assumed to vary linearly with average dose.

Maximum permissible dose rate for general public	5 mSv per year
Maximum permissible dose rate for radiation workers	50 mSv per year
Natural background dose rate	1.25 mSv per year
Dose rate due to industrial, medical, and agricultural use	120 μSv per year
Maximum dose rate due to atmospheric bomb testing (1954–61)	12 μSv per year
Average dose rate due to nuclear reactors (general population, 1980)	2 μSv per year
Threshold for induction of cataract	15 Sv life total

Threshold for nausea		1 Sv in a few hours
Threshold for fatality	There is some	1.5 2 Sv in a few hours
50% fatality within 30 days (from infection)	controversy about	3 Sv in a few hours
Gastro-intestinal death within 3–5 days	these figures because	10 Sv in a few hours
Central nervous system death within hours	no direct evidence.	≈ 20 Sv in a few hours

Reference: American Institute of Physics Handbook (for Tables 2.4 to 2.6).

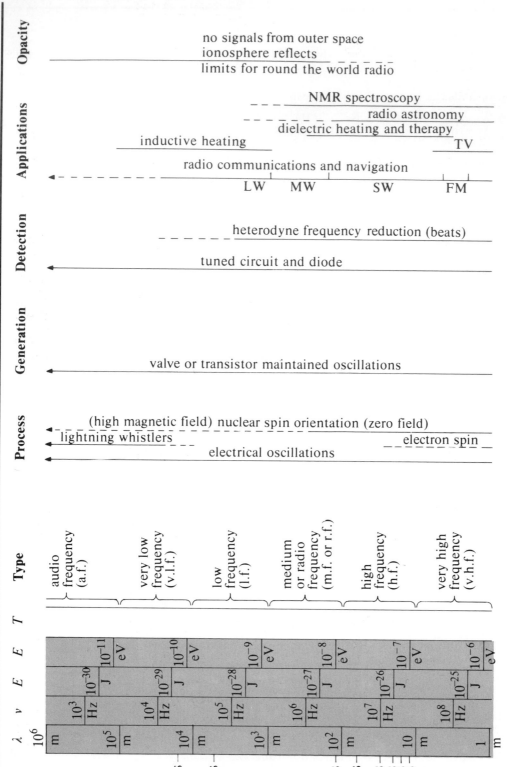

Opacity

quartz opaque

glass opaque

atmosphere opaque – absorption by
O_3 CO_2 H_2O H_2O O_2 H_2O drops

Applications

ESR spectroscopy

radio astronomy

TV   microwave links

radar

radio communication   i.r. molecular absorption spectroscopy

semiconductor devices

Detection

masers

frequency reduction

tuned cavity and diode

tuned circuit   heating effect – bolometer

Generation

masers   globar lamp (SiC at 1200 K)

travelling wave tube   lasers

magnetron

klystron   mercury quartz lamp

electrical oscillations   hot body

Process

molecular rotations

molecular inversions   molecular vibrations

orientation

electrical oscillations   black-body radiation

microwaves

Type

ultra high frequency (u.h.f.)

centimetric super high frequency (s.h.f.)

millimetric extra high frequency (e.h.f.)

sub millimetric or far infra-red (f.i.r.)

intermediate infra-red (i.i.r.)

near infra-red (i.r.)

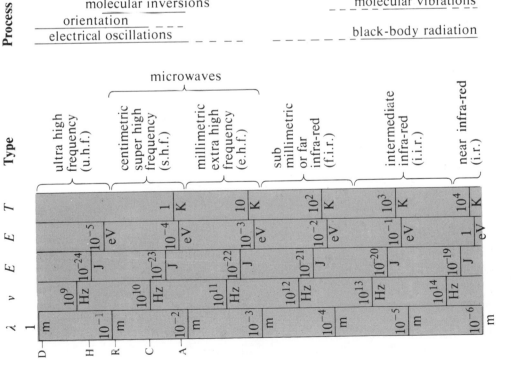

T	1 K	10 K	10^2 K	10^3 K	10^4 K
E	10^{-5} eV	10^{-4} eV	10^{-3} eV	10^{-2} eV	10^{-1} eV	1 eV
E	10^{-24} J	10^{-23} J	10^{-22} J	10^{-21} J	10^{-20} J	10^{-19} J
ν	10^9 Hz	10^{10} Hz	10^{11} Hz	10^{12} Hz	10^{13} Hz	10^{14} Hz
λ	1 m	10^{-1} m	10^{-2} m	10^{-3} m	10^{-4} m	10^{-5} m	10^{-6} m

D   H   R   C   A

λ = wavelength ν = frequency E = quantum energy
T = corresponding temperature = E/k
A NH_3, 23.87 GHz, maser frequency.
C ^{133}Cs, 9193 MHz, definition of second, maser frequency.
D ^{2}D, 327 MHz.   H ^{1}H, 1420 MHz.   R ^{85}Rb, 3036 MHz.
S NBS broadcast standard frequencies, 20, 60 kHz, 2.5, 5, 10, 15, 25 kHz.

3

Opacity

quartz opaque
glass opaque

quartz opaque
glass opaque

atmosphere opaque

atmosphere opaque due to ionization dissociation

Applications

radiant heating
chemical photolysis
photography

X-ray radiography
X-ray crystallography

optical spectroscopy
i.r. spectroscopy
semiconductor devices

u.v. spectroscopy

X-ray spectroscopy

semiconductor

Detection

photo-conduction
photo-electric effect
eye
photographic emulsion

scintillation counter

ionization chamber etc.

heating effect – bolometer

Generation

globar lamp
lasers
arc or spark
gas discharge tube
incandescent body

X-ray tube

Process

atomic inner electron

molecular rotations
molecular vibrations
atomic valence electron transitions
black-body radiation

Bremsstrahlung or
nuclear transitions

Type

far infra-red

intermediate infra-red (i.i.r.)

near infra-red (i.r.)

visible

ultra-violet (u.v.)

vacuum ultra-violet (u.v.)

soft X-rays

T	10^3 K	10^4 K	10^5 K	10^6 K	10^7 K	10^8 K
E	10^{-1} eV	1 eV	10 eV	10^2 eV	10^3 eV	10^4 eV
E	10^{-20} J	10^{-19} J	10^{-18} J	10^{-17} J	10^{-16} J	10^{-15} J
ν	10^{13} Hz	10^{14} Hz	10^{15} Hz	10^{16} Hz	10^{17} Hz	10^{18} Hz
λ	10^{-4} m	10^{-5} m	10^{-6} m	10^{-7} m	10^{-8} m	10^{-9} m 10^{-10} m

K E   H   F   X

Opacity

glass opaque

atmosphere opaque

(superficial) X-ray therapy (deep)

Applications

X-ray radiography

medical   industrial

X-ray crystallography

X-ray spectroscopy

gamma-ray spectroscopy

Detection

detectors

scintillation counter

electron-positron pair production

ionization chamber etc.

photographic emulsion

Generation

radioactive source

X-ray tube

bombardment by accelerated charged particles

Process

transitions

braking radiation

nuclear transitions

fundamental particle reactions and annihilations

black-body radiation   cosmic rays

Type

hard X-rays and gamma rays[1] (γ)

gamma rays (γ)

λ ν E E T

10^9 K	10^{10} K	10^7 eV	10^8 eV	10^9 eV	10^{10} eV
10^5 eV	10^6 eV				
10^{-14} J	10^{-13} J	10^{-12} J	10^{-11} J	10^{-10} J	10^{-9} J
10^{19} Hz	10^{20} Hz	10^{21} Hz	10^{22} Hz	10^{23} Hz	10^{24} Hz
10^{-10} m	10^{-11} m	10^{-12} m	10^{-13} m	10^{-14} m	10^{-15} m	10^{-16} m

P   π   A

ˣ K_α(Cu), 15.4 nm, frequently used for crystallography. ᵖ π^0 meson decay, 68 MeV. [1] X-rays are produced by electron bombardment or deceleration; γ-rays originate in nuclear processes.

ᴬ Proton-antiproton annihilation, 938 MeV. ᴱ Eye peak response, 550 nm. ᴴ H_α line, 122 nm. ᶠ ^{57}Fe, 14.4 keV, Mossbauer line. ᵖ Electron-positron annihilation, 511 keV. ᴷ ^{86}Kr, 605.8 nm, defining metre.

3

See also Colour Table C.2 'Spectra in colour'.

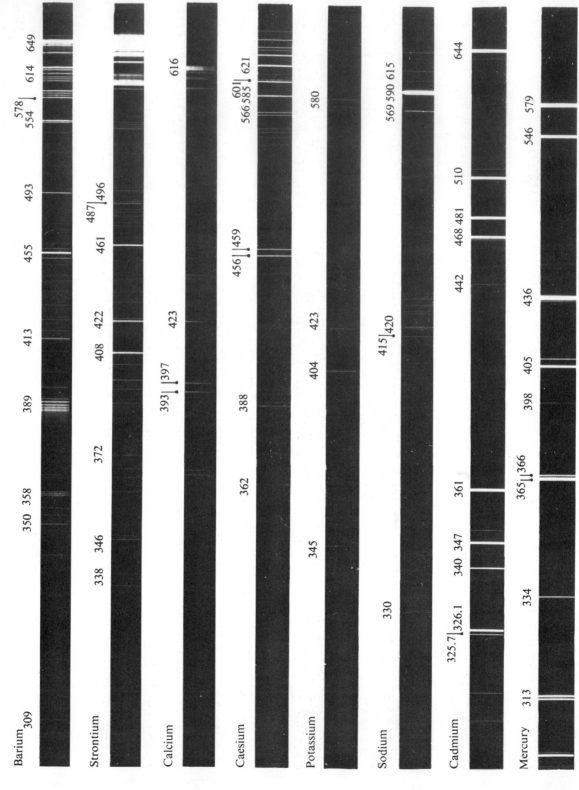

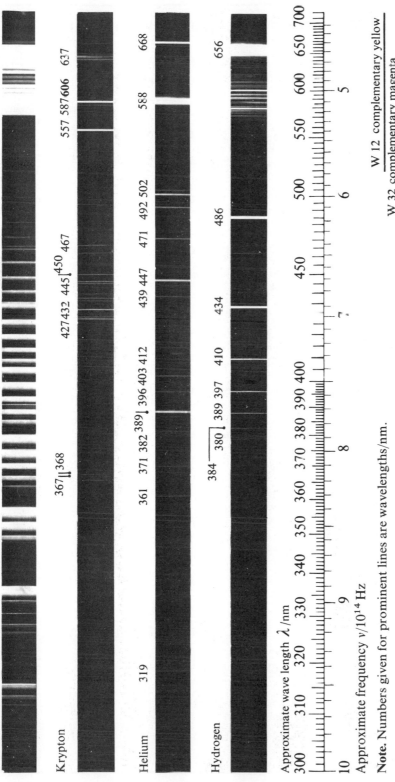

Nitrogen

Krypton

319 367‖368 427 432 445 450 467 557 587 606 637

Helium

361 371 382 389 396 403 412 439 447 471 492 502 588 668

Hydrogen

384 380 389 397 410 434 486 656

Approximate wave length λ/nm

300 310 320 330 340 350 360 370 380 390 400 450 500 550 600 650 700

10 9 8 7 6 5

Approximate frequency ν/10^{14} Hz

Note. Numbers given for prominent lines are wavelengths/nm.

Pass bands of selected filters
(more than 10% of peak transmission)
W Kodak Wratten filter
I Ilford filter
* Mercury monochromat

W 12 complementary yellow
W 32 complementary magenta
W 25 red
W 72B
W 22
W 70, I 609
I 608
W 73, I 606
I 607
W 77*
W 12*
W 58 green
W 74*
I 605
W 44 complementary cyan
I 47 blue
W 75, I 603
I 604
W 76
I 602
W 50*
I 601

Colour (approximate)	← invisible (uv)	purple	dk. bl	blue	bl gn	green	y	o	red	invis

References: Harrison, manufacturers' data.

3

CORRELATION OF INFRA-RED ABSORPTION WAVENUMBERS WITH MOLECULAR STRUCTURE

Intensity w weak absorption; m medium absorption; s strong absorption; v variable intensity of absorption; sh sharp absorption; b broad absorption.

Group	Intensity	Wavenumber range/cm^{-1}
C—H STRETCHING VIBRATIONS		
Alkane	m-s	2962–2853
Alkene	m	3095–3010
Alkyne	s	3300
Arene	v	3030
Aldehyde	w	2900–2820
	and w	2775–2700
C—H BENDING VIBRATIONS		
Alkane	v	1485–1365
Arene		
5 adjacent hydrogen atoms	v, s	750
	and v, s	700
4 adjacent hydrogen atoms	v, s	750
3 adjacent hydrogen atoms	v, m	780
2 adjacent hydrogen atoms	v, m	830
1 isolated hydrogen atom	v, w	880
N—H STRETCHING VIBRATIONS		
Amine, not hydrogen bonded	m	3500–3300
Amide	m	3500–3140
O—H STRETCHING VIBRATIONS		
Alcohols and phenols		
not hydrogen bonded	v, sh	3650–3590
hydrogen bonded	v, b	3750–3200
Carboxylic acids, hydrogen bonded	w	3300–2500
C—O STRETCHING VIBRATIONS		
Esters		
methanoates	s	1200–1180
ethanoates	s	1250–1230
propanoates, etc.	s	1200–1150
benzoates	s	1310–1250
	and s	1150–1100

Group	Intensity	Wavenumber range/cm^{-1}
CARBON-HALOGEN STRETCHING VIBRATIONS		
C—F	s	1400–1000
C—Cl	s	800–600
C—Br	s	600–500
C—I	s	about 500
C=C STRETCHING VIBRATIONS		
Isolated alkene	v	1669–1645
Arene	v	1600
	m	1580
	v	1500
	and m	1450
C=O STRETCHING VIBRATIONS		
Aldehydes, saturated, alkyl	s	1740–1720
Ketones, aryl, alkyl	s	1700–1680
Carboxylic acids		
saturated, alkyl	s	1725–1700
aryl	s	1700–1680
Carboxylic acid anhydrides, saturated	s	1850–1800
	and s	1790–1740
Acyl halides		
chlorides	s	1795
bromides	s	1810
Esters		
saturated	s	1750–1735
alkenyl esters	s	1800–1770
Amides	s	1700–1630
TRIPLE BOND STRETCHING VIBRATIONS		
C≡N	m	2260–2215
C≡C	v, m	2260–2100

INFRA-RED CORRELATION TABLE

Wavenumber range/cm^{-1}	Group
3750–3200	alcohols and phenols, hydrogen bonded
3650–3590	alcohols and phenols, not hydrogen bonded
3500–3300	amine, not hydrogen bonded
3500–3140	amide
3300	alkyne
3095–3010	alkene
3030	arene
2962–2853	alkane
2900–2820	aldehyde
2775–2700	aldehyde
3300–2500	carboxylic acid, hydrogen bonded
2260–2215	C≡N
2260–2100	C≡C
1850–1800	carboxylic acid anhydride
1810	acyl bromide
1800–1770	alkenyl ester
1790–1740	carboxylic acid anhydride
1795	acyl chloride
1750–1735	ester
1740–1720	aldehyde
1730–1717	aryl ester
1725–1700	carboxylic acid
1700–1680	aryl and alkyl ketones, aryl carboxylic acid
1700–1630	amides
1669–1645	alkene
1600, 1580, 1500, and 1450	arene
1485–1365	alkane
1400–1000	fluoroalkane
1310–1250	benzoate
1250–1230	ethanoate
1200–1180	methanoate
1200–1150	propanoate, etc
1150–1100	benzoate
880–700	arene
800–600	chloroalkane
600–500	bromoalkane
about 500	iodoalkane

Reference: Ault.

Chemical shifts for hydrogen relative to TMS (tetramethylsilane) and calculated from the relationship

$$\delta/\text{ppm from TMS} = \frac{B_{\text{TMS}} - B_{\text{sample}}}{B_{\text{TMS}}} \times 10^6 \quad (B \text{ is the applied magnetic flux density}).$$

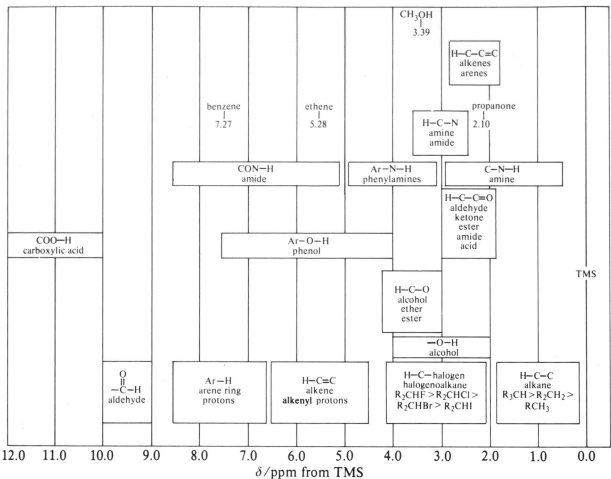

CHEMICAL SHIFTS OF ARENE HYDROGENS FOR SELECTED SUBSTITUTED BENZENES

Chemical shift of ring hydrogens δ/ppm from TMS

Substituent	1, 2-	1, 3-	1, 4-
—NH$_2$	6.5	7.0	6.6
—OH	6.8	7.1	6.9
—Br	7.0	7.4	7.3
—Cl	7.2	7.3	7.3
—CH$_3$	7.1	7.2	7.1
—CH$_2$X	7.3	7.3	7.3
—CHO	7.8	7.5	7.6
—COCH$_3$	7.9	7.4	7.6
—COOH	8.1	7.4	7.5
—NO$_2$	8.2	7.4	7.6

Reference: Doyle.

$K_{\alpha2}$ Emission line for electron transition from L shell to K shell.
$K_{\beta3}$ Emission line for electron transition from M shell to K shell.
L_1 Emission line for electron transition from M shell to L shell.
A Absorption edge for electron removal from K shell.
Z Atomic number. λ Wavelength.

Z	Element	$K_{\alpha2}$ λ/pm	$K_{\beta3}$ λ/pm	L_1 λ/pm	A λ/pm	Z	Element	$K_{\alpha2}$ λ/pm	$K_{\beta3}$ λ/pm	L_1 λ/pm	A λ/pm
19	K	374.5	345.4		344.3	33	As	118.0	105.7	1107.0	104.6
20	Ca	336.2	309.0		307.6	34	Se	110.9	99.2	1029.3	98.1
21	Sc	303.5	278.0		276.3	35	Br	104.4	93.3	958.3	92.2
22	Ti	275.2	251.4		250.2	36	Kr	98.6	88.0		86.7
23	V	250.7	228.4		227.4	37	Rb	93.0	82.9		81.7
24	Cr	229.4	208.5		207.4	38	Sr	87.9	78.3	783.8	77.1
25	Mn	210.6	191.0		190.0	39	Y	83.3	74.1		72.9
26	Fe	194.0	175.7	2016.1	174.7	40	Zr	79.0	70.2	691.3	69.0
27	Co	179.3	162.1	1823.7	161.1	41	Nb	75.0	66.6	652.3	65.4
28	Ni	166.2	150.0	1658.3	149.1	42	Mo	71.4	63.2		62.1
29	Cu	154.4	139.2	1522.1	138.3	74	W	21.4	18.4	167.8	17.8
30	Zn	143.9	129.5	1397.8	128.6	79	Au	18.5	15.9	146.0	15.3
31	Ga	134.4	120.8	1291.6	119.7	82	Pb	17.0	14.6	135.0	14.1
32	Ge	125.8	112.9	1194.4	111.8						

Note: X-ray emission lines are actually close multiplets. The wavelengths given here are for the longer wavelength (and less intense) lines of the K doublets. The L lines are not resolved.

Reference: Siegbahn.

ABSORPTION OF X-RAYS AND γ-RAYS | **3·6**

E Photon energy. ρ Density.
μ_m Mass absorption coefficient for energy absorption (ignoring Compton effect).

Absorber	ρ/g cm^{-3}	E/MeV 0.01	0.05	0.1	0.4 μ_m/cm^2 g^{-1}	1.0	2.0	5.0	10.0
air	0.0013^{273K}	4.55	0.203	0.155	0.095	0.064	0.044	0.027	0.020
water	1.0	4.72	0.221	0.171	0.106	0.070	0.049	0.030	0.022
aluminium	2.7	24.3	0.353	0.169	0.093	0.061	0.043	0.018	0.023
iron	7.9	169	1.90	0.37	0.094	0.060	0.042	0.031	0.030
lead	11.5	150$^{0.02}$	8.5	5.46^A	0.220	0.070	0.046	0.043	0.050
concrete	2.35*	24.6	0.35	0.17	0.095	0.063	0.045	0.029	0.023

^{A} μ_m(Pb) increases discontinuously from 0.95 to 7.2 cm^2 g^{-1} as E increases through the value 0.88 MeV.   * Variable.

Reference: American Institute of Physics Handbook.

$E_{m,j}/\text{kJ mol}^{-1}$

Z		j=1	2	3	4	5	6	7	8	9	10	11	12	13	14
1	H	1312													
2	He	2372	5251												
3	Li	520	7298	11815											
4	Be	900	1757	14849	21007										
5	B	801	2427	3660	25026	32828									
6	C	1086	2353	4621	6223	37832	47278								
7	N	1402	2856	4578	7475	9445	53268	64362							
8	O	1314	3388	5301	7469	10989	13327	71337	84080						
9	F	1681	3374	6051	8408	11022	15164	17868	92040	106437					
10	Ne	2081	3952	6122	9370	12177	15239	19999	23069	115382	131435				
11	Na	496	4563	6913	9544	13352	16611	20115	25491	28934	141367	159079			
12	Mg	738	1451	7733	10541	13629	17995	21704	25657	31644	35463	169996	189371		
13	Al	578	1817	2745	11578	14831	18378	23296	27460	31862	38458	42655	201276	222313	
14	Si	789	1577	3232	4356	16091	19785	23787	29253	33878	38734	45935	50512	235211	257928
15	P	1012	1903	2912	4957	6274	21269	25398	29855	35868	40960	46274	54074	59037	271807
16	S	1000	2251	3361	4564	7012	8496	27107	31671	36579	43140	48706	54483	62876	68232
17	Cl	1251	2297	3822	5158	6542	9362	11018	33606	38601	43963	51068	57119	63364	72342
18	Ar	1521	2666	3931	5771	7238	8781	11996	13842	40761	46188	52003	59654	66201	72920
19	K	419	3051	4412	5877	7975	9649	11343	14942	16964	48577	54433	60701	68896	75950
20	Ca	590	1145	4912	6474	8144	10496	12320	14207	18192	20385	57050			
21	Sc	631	1235	2389	7089	8844	10720	13320	15313	17370	21741	24106			
22	Ti	658	1310	2653	4175	9573	11517	13586	16259	18640	20833	25592			
23	V	650	1414	2828	4507	6294	12362	14490	16760	19860	22240	24609			
24	Cr	653	1592	2987	4740	6686	8738	15540	17822	20200	23580	26130			
25	Mn	717	1509	3249	4940	6985	9200	11508	18956	21400	23960	27600			
26	Fe	759	1561	2958	5290	7236	9600	12100	14576	22679	25290	28020			
27	Co	758	1646	3232	4950	7671	9840	12400	15100	17960	26600	29400			
28	Ni	737	1753	3394	5300	7285	10400	12800	15600	18600	21660	30990			
29	Cu	746	1958	3554	5330	7709	9940	13400	16000	19200	22400	25700			
30	Zn	906	1733	3833	5730	7970	10400	12900	16800	19600	23000	26400			
31	Ga	579	1979	2963	6200										
32	Ge	762	1537	3302	4411	9021									
33	As	947	1798	2736	4837	6043	12312								
34	Se	941	2045	2974	4144	6590	7883	14990							
35	Br	1140	2100	3500	4560	5760	8549	9938	18600						

See below for j = 15 to 19.

Successive molar ionization energies E_{mj} / kJ mol^{-1}

Z		1	2	3	4	5	6	7	8	9	10	11
36	Kr	1351	2368	3565	5070	6243	7574	10710	12158	22230		
37	Rb	403	2632	3900	5080	6850	8144	9572	13100	14500	26740	31270
38	Sr	550	1064	4210	5500	6908	8761	10200	11800	15600	17100	19900
39	Y	616	1181	1980	5960	7429	8973	11200	12400	14137	18400	
40	Zr	660	1267	2218	3313	7863	9500					
41	Nb	664	1382	2416	3700	4877	9900	12100				
42	Mo	685	1558	2621	4480	5905	6600	12230	14800			
43	Tc	702	1472	2850								
44	Ru	711	1617	2747								
45	Rh	720	1745	2997								
46	Pd	805	1875	3177								
47	Ag	731	2074	3361								
48	Cd	868	1631	3616								
49	In	558	1821	2705	5200							
50	Sn	709	1412	2943	3930	6974						
51	Sb	834	1595	2440	4260	5403	10400					
52	Te	869	1790	2698	3610	5668	6820	13200				
53	I	1008	1846	3200								
54	Xe	1170	2047	3100								
55	Cs	376	2420	3300						16400		
56	Ba	503	965									
57	La	538	1067	1850	4820							
63	Eu	547	1085	2404	4110							
72	Hf	680	1440	2250	3215							
73	Ta	761										
74	W	770	1700									
75	Re	760	1600									
76	Os	840	1600									
77	Ir	880										
78	Pt	870	1791									
79	Au	890	1980	3300								
80	Hg	1007	1810									
81	Tl	589	1971	2878	4900							
82	Pb	716	1450	3082	4083	6638	8500					
83	Bi	703	1610	2466	4370	5403						
84	Po	812										
86	Rn	1037										

Z		15	16	17	18	19
15	P	296201				
16	S	311068	337146			
17	Cl	78098	353001	380768		
18	Ar	82475	88578	397614	427076	
19	K	83152	93403	99771	444911	476075

DEFINITION

E_{mj} Successive molar ionization energy, that is, the energy needed to remove the jth successive electron from the atoms or ions.

Z Atomic number of the element.

Atomic ionization energies E_j (that is, for single atoms/ions), ionization potentials V_j, corresponding frequencies v, and corresponding wavenumbers σ, are related to E_{mj} by

$$E_j = E_{mj}/L; \quad E_j/J = 1.66 \times 10^{-21} E_{mj}/\text{kJ mol}^{-1} \text{ and}$$
$$E_j/\text{eV} = 1.04 \times 10^{-2} E_{mj}/\text{kJ mol}^{-1}$$
$$V_j = E_{mj}/F; \quad V_j/V = 1.04 \times 10^{-2} E_{mj}/\text{kJ mol}^{-1}$$
$$v = E_{mj}/Lh; \quad v/\text{Hz} = 2.50 \times 10^{12} E_{mj}/\text{kJ mol}^{-1}$$
$$\sigma = E_{mj}/Lhc; \quad \sigma/\text{cm}^{-1} = 83.5\, E_{mj}/\text{kJ mol}^{-1}$$

Reference: Moore.

4

This table gives the electronic configurations of the elements in their ground states.

Shell subshell		K 1s	L 2s	2p	M 3s	3p	3d	N 4s	4p	4d	4f	O 5s	5p	5d	5f	5g
1	H	1														
2	He	2														
3	Li	2	1													
4	Be	2	2													
5	B	2	2	1												
6	C	2	2	2												
7	N	2	2	3												
8	O	2	2	4												
9	F	2	2	5												
10	Ne	2	2	6												
11	Na	2	2	6	1											
12	Mg	2	2	6	2											
13	Al	2	2	6	2	1										
14	Si	2	2	6	2	2										
15	P	2	2	6	2	3										
16	S	2	2	6	2	4										
17	Cl	2	2	6	2	5										
18	Ar	2	2	6	2	6										
19	K	2	2	6	2	6		1								
20	Ca	2	2	6	2	6		2								
21	Sc	2	2	6	2	6	1	2								
22	Ti	2	2	6	2	6	2	2								
23	V	2	2	6	2	6	3	2								
24	Cr	2	2	6	2	6	5	1								
25	Mn	2	2	6	2	6	5	2								
26	Fe	2	2	6	2	6	6	2								
27	Co	2	2	6	2	6	7	2								
28	Ni	2	2	6	2	6	8	2								
29	Cu	2	2	6	2	6	10	1								
30	Zn	2	2	6	2	6	10	2								
31	Ga	2	2	6	2	6	10	2	1							
32	Ge	2	2	6	2	6	10	2	2							
33	As	2	2	6	2	6	10	2	3							
34	Se	2	2	6	2	6	10	2	4							
35	Br	2	2	6	2	6	10	2	5							
36	Kr	2	2	6	2	6	10	2	6							
37	Rb	2	2	6	2	6	10	2	6			1				
38	Sr	2	2	6	2	6	10	2	6			2				
39	Y	2	2	6	2	6	10	2	6	1		2				
40	Zr	2	2	6	2	6	10	2	6	2		2				
41	Nb	2	2	6	2	6	10	2	6	4		1				
42	Mo	2	2	6	2	6	10	2	6	5		1				
43	Tc	2	2	6	2	6	10	2	6	6		1				
44	Ru	2	2	6	2	6	10	2	6	7		1				
45	Rh	2	2	6	2	6	10	2	6	8		1				
46	Pd	2	2	6	2	6	10	2	6	10						
47	Ag	2	2	6	2	6	10	2	6	10		1				
48	Cd	2	2	6	2	6	10	2	6	10		2				
49	In	2	2	6	2	6	10	2	6	10		2	1			
40	Sn	2	2	6	2	6	10	2	6	10		2	2			
51	Sb	2	2	6	2	6	10	2	6	10		2	3			
52	Te	2	2	6	2	6	10	2	6	10		2	4			
53	I	2	2	6	2	6	10	2	6	10		2	5			
54	Xe	2	2	6	2	6	10	2	6	10		2	6			

D-block elements (shaded: 21–30, 39–48)

Shell subshell		K	L	M	N				O					P				Q
					4s	4p	4d	4f	5s	5p	5d	5f	5g	6s	6p	6d	6(f, g, h)	7s
55	Cs	2	8	18	2	6	10		2	6				1				
56	Ba	2	8	18	2	6	10		2	6				2				
57	La	2	8	18	2	6	10		2	6	1			2				
58	Ce	2	8	18	2	6	10	2	2	6				2				
59	Pr	2	8	18	2	6	10	3	2	6				2				
60	Nd	2	8	18	2	6	10	4	2	6				2				
61	Pm	2	8	18	2	6	10	5	2	6				2				
62	Sm	2	8	18	2	6	10	6	2	6				2				
63	Eu	2	8	18	2	6	10	7	2	6				2				
64	Gd	2	8	18	2	6	10	7	2	6	1			2				
65	Tb	2	8	18	2	6	10	9	2	6				2				
66	Dy	2	8	18	2	6	10	10	2	6				2				
67	Ho	2	8	18	2	6	10	11	2	6				2				
68	Er	2	8	18	2	6	10	12	2	6				2				
69	Tm	2	8	18	2	6	10	13	2	6				2				
70	Yb	2	8	18	2	6	10	14	2	6				2				
71	Lu	2	8	18	2	6	10	14	2	6	1			2				

Lanthanides

Shell subshell		K	L	M	4s	4p	4d	4f	5s	5p	5d	5f	5g	6s	6p	6d	6(f, g, h)	7s
72	Hf	2	8	18	2	6	10	14	2	6	2			2				
73	Ta	2	8	18	2	6	10	14	2	6	3			2				
74	W	2	8	18	2	6	10	14	2	6	4			2				
75	Re	2	8	18	2	6	10	14	2	6	5			2				
76	Os	2	8	18	2	6	10	14	2	6	6			2				
77	Ir	2	8	18	2	6	10	14	2	6	7			2				
78	Pt	2	8	18	2	6	10	14	2	6	9			1				
79	Au	2	8	18	2	6	10	14	2	6	10			1				
80	Hg	2	8	18	2	6	10	14	2	6	10			2				

D-block elements

Shell subshell		K	L	M	4s	4p	4d	4f	5s	5p	5d	5f	5g	6s	6p	6d	6(f, g, h)	7s
81	Tl	2	8	18	2	6	10	14	2	6	10			2	1			
82	Pb	2	8	18	2	6	10	14	2	6	10			2	2			
83	Bi	2	8	18	2	6	10	14	2	6	10			2	3			
84	Po	2	8	18	2	6	10	14	2	6	10			2	4			
85	At	2	8	18	2	6	10	14	2	6	10			2	5			
86	Rn	2	8	18	2	6	10	14	2	6	10			2	6			
87	Fr	2	8	18	2	6	10	14	2	6	10			2	6			1
88	Ra	2	8	18	2	6	10	14	2	6	10			2	6			2
89	Ac	2	8	18	2	6	10	14	2	6	10			2	6	1		2
90	Th	2	8	18	2	6	10	14	2	6	10			2	6	2		2
91	Pa	2	8	18	2	6	10	14	2	6	10	2		2	6	1		2
92	U	2	8	18	2	6	10	14	2	6	10	3		2	6	1		2
93	Np	2	8	18	2	6	10	14	2	6	10	4		2	6	1		2
94	Pu	2	8	18	2	6	10	14	2	6	10	6		2	6			2
95	Am	2	8	18	2	6	10	14	2	6	10	7		2	6			2
96	Cm	2	8	18	2	6	10	14	2	6	10	7		2	6	1		2
97	Bk	2	8	18	2	6	10	14	2	6	10	9		2	6			2
98	Cf	2	8	18	2	6	10	14	2	6	10	10		2	6			2
99	Es	2	8	18	2	6	10	14	2	6	10	11		2	6			2
100	Fm	2	8	18	2	6	10	14	2	6	10	12		2	6			2
101	Md	2	8	18	2	6	10	14	2	6	10	13		2	6			2
102	No	2	8	18	2	6	10	14	2	6	10	14		2	6			2
103	Lr	2	8	18	2	6	10	14	2	6	10	14		2	6	1		2

Actinides

Shell subshell		K	L	M	4s	4p	4d	4f	5s	5p	5d	5f	5g	6s	6p	6d	6(f, g, h)	7s
104	Rf	2	8	18	2	6	10	14	2	6	10	14		2	6	2		2
105	Ha	2	8	18	2	6	10	14	2	6	10	14		2	6	3		2
106		2	8	18	2	6	10	14	2	6	10	14		2	6	4		2

Beyond $_{94}$Pu, the assignments are conjectural.

This table gives *some* of the energy levels available to atoms of elements with one or two outer electrons. The ground state is printed in bold. Transitions between the levels give rise to infra-red, visible, and ultra-violet spectrum lines. Except in the case of hydrogen, transitions normally take place only between a level printed in upright type and a level printed in sloping type. In the case of the elements with two outer electrons, the energy levels occur in two sets marked S and T. Transitions do not normally take place between the levels in these two sets except between the lowest S and the lowest T levels.

Figures give the value of $E/aJ = E/10^{-18}$ J, where E is the difference between the energy of a neutral atom in the quantized state and the composite energy of an ionized atom in its lowest energy state and an electron at rest well away from the atom. Values are known to an accuracy of 1 in 10^5, but they are given here to three decimal places only.

Elements with one outer electron

Hydrogen	Lithium	Sodium	Potassium	Rubidium
−0.022	−0.087	−0.082 [D]	−0.079 [D]	−0.068
−0.027	−0.095	−0.088	−0.095	−0.099
−0.034	−0.103	−0.101	−0.096	−0.117 [D]
−0.044	−0.136	−0.127 [D]	−0.119 [D]	−0.147
−0.061	−0.139	−0.137	−0.150	−0.158 [D]
−0.087	−0.168	−0.164	−0.151 [D]	−0.198 [D]
−0.136	−0.242	−0.222 [D]	−0.205 [D]	−0.269
−0.242	−0.250	−0.244	−0.268 [D]	−0.285
−0.545	−0.323	−0.312	−0.278	−0.419 [D]
−2.180	−0.568	−0.487 [D]	−0.437 [D]	**−0.669**
	−0.864	**−0.823**	**−0.695**	

Elements with two outer electrons

Helium S	Helium T	Beryllium S	Beryllium T	Magnesium S	Magnesium T
−0.136	−0.136	−0.127	−0.144	−0.106	−0.093
−0.136	−0.141	−0.214	−0.212	−0.138	−0.115
−0.146	−0.159	−0.407	−0.261	−0.169	−0.147
−0.240	−0.243	−0.648	−0.459	−0.181	−0.149
−0.243	−0.254	**−1.494**	−1.057 [T]	−0.245	−0.195
−0.267	−0.300			−0.303	−0.272
−0.540	−0.580			−0.361	−0.275 [T]
−0.636	−0.764			−0.529	−0.407
−1.953				**−1.225**	−0.791 [T]

Calcium S	Calcium T	Strontium S	Strontium T	Barium S	Barium T
−0.149	−0.106 [T]	−0.139	−0.103	−0.100	−0.098
−0.151	−0.135	−0.149	−0.169	−0.188	−0.161
−0.238	−0.175	−0.221	−0.217 [T]	−0.274	−0.224 [T]
−0.250	−0.230 [T]	−0.235	−0.241 [T]	−0.326	−0.225 [T]
−0.318	−0.253 [T]	−0.305	−0.336	−0.476	−0.315
−0.509	−0.353	−0.481	−0.552 [T]	−0.609	−0.591 [T]
−0.545	−0.575 [T]	−0.512	−0.628 [T]	**−0.835**	−0.655 [T]
−0.979	−0.678 [T]	**−0.912**			

Zinc S	Zinc T	Cadmium S	Cadmium T	Mercury S	Mercury T
−0.193	−0.205	−0.188	−0.198	−0.194	−0.203
−0.255	−0.258 [T]	−0.251	−0.259 [T]	−0.255	−0.255 [T]
−0.264	−0.288 [T]	−0.265	−0.287 [T]	−0.256	−0.291 [T]
−0.396	−0.439	−0.381	−0.418	−0.402	−0.434
−0.576	−0.863 [T]	−0.573	−0.842 [T]	−0.598	−0.924 [T]
−1.505		**−1.440**		**−1.672**	

Note

For those familiar with spectroscopic notation, the S and T sets are singlet and triplet states. Levels in sloping type are P states, those in upright type S or D states. No F states and no 'displaced' states have been included, so that a number of prominent spectral lines cannot be obtained from this table.

[D] This level is the lower of the two closely spaced levels (giving rise to spectral doublets on transition).

[T] This level is the lower of three closely spaced levels (giving rise to spectral triplets on transition).

Reference: Kuhn.

This table is arranged according to group in the Periodic Table.

r_V Van der Waals radius.
r_m Metallic radius for coordination number 12[A].
r_{cov} Covalent radius.
r_i Ionic radius for coordination number 6[B], except where superscript number ^{4} ^{3} etc. indicates different coordination number (figures in parentheses give charge state).
N_P Pauling electronegativity index[C].

Group		r_V/nm	r_m/nm	r_{cov}/nm	r_i/nm	N_P
I	Li	0.18	0.157	0.134	0.074(+1)	1.0
	Na	0.230	0.191	0.154	0.102(+1)	0.9
	K	0.280	0.235	0.196	0.138(+1)	0.8
	Rb		0.250		0.149(+1)	0.8
	Cs		0.272		0.170(+1)	0.7
	NH₄⁺				0.150(+1)	
II	Be		0.112	0.090	$0.027(+2)^{4}$	1.5
	Mg	0.170	0.160	0.145	0.072(+2)	1.2
	Ca		0.197		0.100(+2)	1.0
	Sr		0.215		0.113(+2)	1.0
	Ba		0.224		0.136(+2)	0.9
III	B		0.098	0.090	$0.012(+3)^{4}$	2.0
	Al		0.143	0.130	0.053(+3)	1.5
	Ga	0.190	0.153	0.12	0.062(+3)	1.6
D block	Sc		0.164		0.075(+3)	1.3
	Ti		0.147		0.086(+2); 0.067(+3); 0.061(+4)	1.5
	V		0.135		0.064(+3)[D]; 0.059(+4); 0.054(+5)	1.6
	Cr		0.129		0.062(+3); $0.030(+6)^{4}$	1.6
	Mn		0.137	0.139	0.067(+2)[E]; 0.053(+3)[E]; $0.026(+7)^{4}$	1.5
	Fe		0.126	0.125	0.061(+2)[E]; 0.055(+3)[E]	1.8
	Co		0.125	0.126	0.065(+2)[E]; 0.053(+3)[E]	1.8
	Ni		0.125	0.121	0.070(+2); 0.056(+3)[E]	1.8
	Cu		0.128	0.135	0.073(+2); $0.046(+1)^{2}$	1.9
	Zn		0.137	0.12	0.075(+2)	1.6
	Mo		0.140		0.060(+6); 0.067(+3)	1.8
	Ag		0.144	0.152	0.115(+1); 0.089(+2); $0.065(+3)^{4}$	1.9

Group		r_V/nm	r_m/nm	r_{cov}/nm	r_i/nm	N_P
	Cd	0.170	0.152	0.148	0.095(+2)	1.7
	Au	0.175	0.144		$0.070(+3)^{4}$; 0.137(+1)	2.4
	Hg	0.170	0.155	0.148	0.102(+2); $0.097(+1)^{3}$	1.9
IV	C	0.17	0.092	0.077		2.5
	Si	0.210	0.132	0.118	0.040(+4)	1.8
	Ge		0.139	0.122	0.054(+4)	1.8
	Sn	0.190	0.158	0.140	0.069(+4); $0.122(+2)^{8}$	1.8
	Pb	0.200	0.175		0.078(+4); 0.118(+2)	1.8
V	N	0.155	0.088	0.075	0.171(−3); $0.012(+5)^{3}$	3.0
	P	0.185	0.128	0.110	0.190(−3); $0.017(+5)^{4}$	2.1
	As		0.139	0.122	0.220(−3); 0.050(+5)	2.0
	Sb		0.161	0.143	$0.080(+3)^{5}$; 0.061(+5)	1.9
	Bi	0.200	0.182		0.102(+3)	1.9
VI	O	0.150	0.089	0.073	0.140(−2)	3.5
	S	0.180	0.127	0.102	0.185(−2); $0.030(+6)^{4}$	2.5
	Se	0.190	0.140	0.117	0.195(−2); $0.050(+6)^{4}$	2.4
	Te	0.210	0.143	0.135	0.220(−2); $0.056(+4)^{3}$	2.1
VII	H	0.12	0.078	0.037	0.208(−1)	2.1
	F	0.155		0.071	0.133(−1)	4.0
	Cl	0.180		0.099	0.180(−1); $0.020(+7)^{5}$	3.0
	Br	0.190		0.114	0.195(−1); $0.026(+7)^{4}$	2.8
	I	0.195		0.133	0.215(−1); 0.095(+5)	2.5
VIII	He	0.18				
	Ne	0.160				
	Ar			0.190		
	Kr				0.200	
	Xe				0.220	

	He	Ne	Ar	Kr	Xe	CH₃	C₆H₆
r_V/nm	0.18	0.160	0.190	0.200	0.220	0.20	0.185

[A] The metallic radius for other coordination numbers (CN) varies according to the following formula: $r_m^{CN=n} = k r_m^{CN=12}$

CN	12	8	6	4
k	1.00	0.97	0.96	(0.88)

[B] These ionic radii are applicable to oxides and fluorides (based on the oxygen ion O²⁻ radius of 0.140 nm). The ionic radius for other coordination numbers varies according to the following formula: $r_i^{CN=n} = k r_i^{CN=6}$

CN	4	6	8
k	0.95	1.00	1.04

[C] The Pauling electronegativity index is only one of a number of electronegativity indexes (Mulliken, Sanderson, Allred-Rochow, etc.). N_P is a measure of how strongly the atom attracts electrons. The percentage of ionic bonding, P, in a bond depends on the difference ΔN_P, in the N_P values of the atoms as follows:

ΔN_P	0.1	0.3	0.5	0.7	1.0	1.3	1.5	1.7	2.0	2.5	3.0	3.2
P/%	0.5	2	6	12	22	34	43	51	63	79	89	92

It can also be used to calculate the bond length from the sum of the covalent radii using the empirical formula: $r_{AB} = r_A + r_B - 0.09|N_A - N_B|$

[D] r_i for V²⁺ is 0.079 nm.   [E] Low spin value.

References: r_V Bondi; r_m Wells, Teatum; r_{cov} Dunod, Pauling; r_i Shannon, Pauling; N_P Pauling.

Linear

$Ag(CN)_2^-$ $BeCl_2$ HCN   HCCH   NCCN CO_2
CS_2 HgX_2 XHgHgX   (X = Br, Cl) I_3^- N_3^-
NNO XeF_2

Bent

$ClO_2(117°)$ $ClO_2^-(111°)$ $ONCl(116°)$ $NO_2(134°)$
$NO_2^-(115°)$ $(CH_3)O(CH_3)(111.5°)$ $Cl_2O(110°)$
$F_2O(103°)$ $H_2O(104.5°)$ $O_3(117°)$ $SCl_2(100°)$
$SO_2(120°)$ $H_2S(92.2°)$ $H_2Se(91°)$
$SnX_2(X = Br, Cl, I)$

T-shape

$BrF_3(86°)$ $ClF_3(89°)$

Trigonal planar (120°)

BX_3 (X = Br, Cl, F) H_2CO
Cl_2CO (ClCCl = 111.3°) CO_3^{2-} GaI_3
NO_3^- SO_3 C_2H_4

Trigonal pyramidal

$AsBr_3(100°)$ $AsCl_3(98°)$ $AsH_3(92°)$ $BiBr_3(100°)$
$BiCl_3(100°)$ $ClO_3^-(110°)$ $N(CH_3)_3(108°)$
$NH_3(107°)$ $PBr_3(101°)$ $PCl_3(100.1°)$ $PH_3(94°)$
SO_3^{2-} $SbBr_3(97°)$ $SbCl_3(100°)$ $SbH_3(91°)$ XeO_3

Trigonal bipyramidal (90° and 120°)

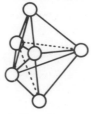

AsF_5 $MoCl_5$ PCl_5 PF_5 $SbCl_5$

Square planar

$AuCl_4^-$ ICl_4^- $Ni(CN)_4^{2-}$ $PdCl_4^{2-}$ $Pt(NH_3)_4^{2+}$
XeF_4 $[Cu(H_2O)_4]^{2+}$

Tetrahedral (109.5°)

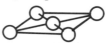

$AlCl_4^-$ BF_4^- BH_4^- CCl_4 ClO_4^- CrO_4^{2-}
$Cu(CN)_4^{3-}$ GeH_4 MnO_4^- NH_4^+ $Ni(CO)_4$
PH_4^+ PO_4^{3-} SO_4^{2-} SeO_4^{2-} $SiCl_4$ $SnCl_4$
$TiCl_4$ VCl_4 $Zn(CN)_4^{2-}$

Octahedral (90°)

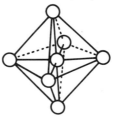

MoF_6 PCl_6^- SF_6 SeF_6 TeF_6 UF_6 WCl_6
XeF_6 (also 6-coordinated complexes of many
metals, e.g. $[Al(H_2O)_6]^{3+}$)

Reference: Aylward.

L Bond length.

$E(X—Y)$ Bond energy, defined:

(a) for X_2 molecules as the molar enthalpy change for the process $X_2(g) \rightarrow 2X(g)$;

(b) for XY_n molecules as the molar enthalpy change for the process

$$\frac{1}{n}XY_n(g) \rightarrow \frac{1}{n}X(g) + Y(g).$$

(Both these processes are at 298 K with individual species pressures of 1 atm.)

Average bond energies for organic compounds are calculated using average values for CH_2 chain increments and alkane-compound increments in $\Delta H_f^\ominus$ (298 K, 1 atm).

	Bond	in	L / nm	$E(X—Y)$ / $kJ\,mol^{-1}$		Bond	in	L / nm	$E(X—Y)$ / $kJ\,mol^{-1}$
1	Br—Br	Br_2	0.228	192.9	33	O—Si	$SiO_2(s)$	0.161	466
2	Br—H	HBr	0.141	366.3	34	O=Si	$SiO_2(g)$	—	638
3	Cl—Cl	Cl_2	0.199	243.4	35	O≡Si	SiO	—	805
4	Cl—H	HCl	0.127	432.0	36	P—P	P_4	0.221	198
5	F—F	F_2	0.142	158	37	P≡P	P_2	0.189	485
6	F—H	HF	0.092	568.0	38	C—C	average	0.154	347
7	I—I	I_2	0.267	151.2	39	C=C	average	0.134	612
8	H—I	HI	0.161	298.3	40	C≡C	average	0.120	838
9	H—H	H_2	0.074	435.9	41	C—H	average	0.108	413
10	H—Si	SiH_4	0.148	318	42	C—H	CH_4	0.109	435
11	H—Ge	GeH_4	0.153	285	43	C—F	average	0.138	467
12	H—N	NH_3	0.101	391	44	C—F	CH_3F	0.139	452
13	H—P	PH_3	0.144	321	45	C—F	CF_4	0.132	485
14	H—As	AsH_3	0.152	297	46	C—Cl	average	0.177	346
15	H—O	H_2O	0.096	464	47	C—Cl	CCl_4	0.177	327
16	H—S	H_2S	0.134	364	48	C⋯Cl	C_6H_5Cl	0.170	—
17	H—Se	H_2Se	0.146	313	49	C—Br	average	0.194	290
18	Na—Na	Na_2	0.308	72	50	C—Br	CBr_4	0.194	285
19	K—K	K_2	0.392	49	51	C—I	average	0.214	228
20	N—N	N_2H_4	0.145	158	52	C—I	CH_3I	0.214	234
21	N=N	$C_6H_{14}N_2$	0.120	410	53	C—N	average	0.147	286
22	N≡N	N_2	0.110	945.4	54	C=N	average	0.130	615
23	N—O	HNO_2	0.120	214	55	C≡N	average	0.116	887
24	N=O	NOF, NOCl	0.114	587	56	C⋯N	phenylamine	0.135	—
25	N≡P	PN	0.149	582	57	C—O	average	0.143	358
26	O—O	H_2O_2	0.148	144	58	C—O	CH_3OH	0.143	336
27	O—O	O_3	0.128	302	59	C=O	CO_2	0.116	805
28	O=O	O_2	0.121	498.3	60	C=O	HCHO	0.121	695
29	S—S	S_8	0.205	266	61	C=O	aldehydes	0.122	736
30	S=S	S_2	0.189	429.2	62	C=O	ketones	0.122	749
31	O—S	SO_3	0.143	469	63	C≡O	CO	0.113	1077
32	Si—Si	Si_2H_6, Si_3H_8	0.235	226	64	C—Si	$(CH_3)_4Si$, SiC(s)	0.187	307

References: Sutton, Johnson, Cottrell.

Compound	Angle/°	Sequence	Length/nm	Bond
CCl_4	109.5	Cl—C—Cl	0.177	Cl—C
CH_4	109.5	H—C—H	0.109	H—C
CH_3Cl	110.5	H—C—H	0.110	H—C
	108.0	Cl—C—H	0.178	Cl—C
CH_2Cl_2	112.0	H—C—H	0.107	H—C
	111.8	Cl—C—Cl	0.177	Cl—C
$CHCl_3$	110.9	Cl—C—Cl	0.107	C—H
			0.176	Cl—C
C_2H_4	117.3	H—C—H	0.109	H—C
			0.134	C—C
C_3H_6 cyclopropane	120.0	H—C—H	0.153	C—C
	120.0	H—C—C	0.107	C—H
C_6H_6 benzene	120.0	C—C—C	0.1084	C—H
			0.1397	C—C
CH_3OH	109	C—O—H	0.143	C—O
			0.096	O—H
(acetic acid)	122.0	O—C—O'	0.131	C—O
	119.5	C—C—O'	0.125	C—O'
	116.0	C—C—O	0.095	O—H
	106.8	H—C—H	0.108	H—C
CH_3CHO	123.9	C—C—O	0.109	H—C
	108.3	H—C—H	0.150	C—C
			0.122	C—O
$(CH_3)_2O$	111.5	C—O—C	0.142	C—O
CH_3NH_2	109.5	H—C—H	Methyl axis makes angle 3.5° with C—N axis	
	112.2	H—N—C	0.109	H—C
	105.8	H—N—H		
$(CH_3)_2NH$	111.0	C—N—C	0.108	H—C
			0.146	C—N
$(CH_3)_3N$	108.7	C—N—C	0.147	C—N
	107.1	H—C—H	0.109	H—C
CO_3^{2-}	120.0	O—C—O	0.129	O—C
$COCl_2$	111.3	Cl—C—Cl	0.175	Cl—C
H_2O	104.5	H—O—H	0.096	H—O
H_2S	92.2	H—S—H	0.134	H—S
H_2Se	91.0	H—Se—H	0.146	H—Se
H_2Te	89.5	H—Te—H	0.17	H—Te
NH_3	107.0	H—N—H	0.101	H—N
NO_2	134.0	O—N—O	0.120	N—O
NO_3^-	120.0	O—N—O	0.124	N—O
PCl_3	100.1	Cl—P—Cl	0.204	Cl—P
PCl_5	120.0	Cl—P—Cl	0.204	Cl—P
	90.0	Cl—P—Cl	0.219	Cl—P
SF_6	90.0	F—S—F	0.156	F—S
SO_3	120.0	O—S—O	0.143	S—O

The CH_3C entry shows the acetic acid structure with =O' at top and OH at bottom.

References: Sutton, Cottrell.

CRYSTAL SYSTEMS

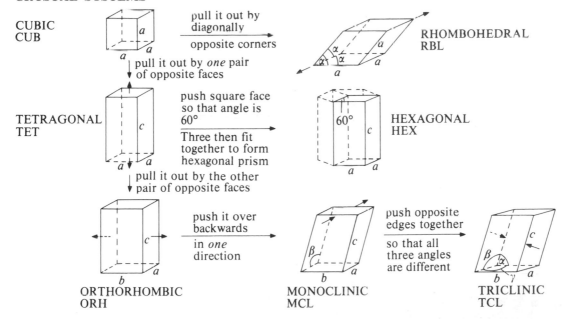

CUBIC
CUB

pull it out by
diagonally
opposite corners

RHOMBOHEDRAL
RBL

pull it out by *one* pair
of opposite faces

TETRAGONAL
TET

push square face
so that angle is
60°

Three then fit
together to form
hexagonal prism

HEXAGONAL
HEX

pull it out by the other
pair of opposite faces

push it over
backwards
in *one*
direction

push opposite
edges together

so that all
three angles
are different

ORTHORHOMBIC
ORH

MONOCLINIC
MCL

TRICLINIC
TCL

4

CRYSTAL STRUCTURES (see also colour tables C.1 and C.3)

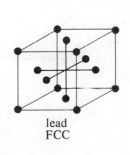

lead
FCC

zinc
HCP

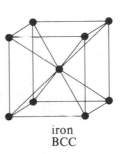

iron
BCC

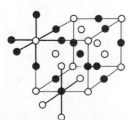

sodium chloride (NaCl)
6:6 coordination
● Na ○ Cl

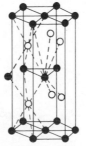

nickel arsenide (NiAs)
6:6 coordination
● As ○ Ni

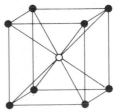

caesium chloride (CsCl)
8:8 coordination
● Cl ○ Cs

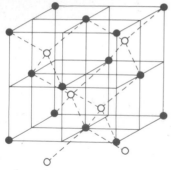

zinc blende (ZnS)
4:4 coordination
● S ○ Zn

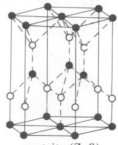

wurtzite (ZnS)
4:4 coordination
● S ○ Zn

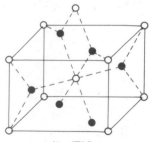

rutile (TiO$_2$)
6:3 coordination
● O ○ Ti

diamond
4 coordination

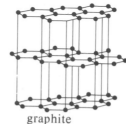

graphite
layer structure

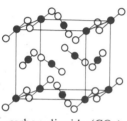

fluorite (CaF$_2$)
8:4 coordination
● Ca ○ F

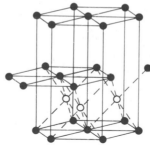

cadmium iodide (CdI$_2$)
layer structure
● I ○ Cd

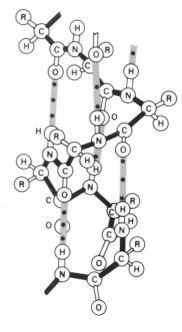

α-helix

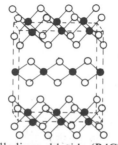

carbon dioxide (CO$_2$)
molecular structure
● C ○ O

palladium chloride (PdCl$_2$)
chain structure
● Pd ○ Cl

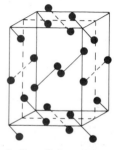

iodine (I$_2$)
molecular structure

Zinc blende, ZnS

Marcasite, FeS$_2$

Calcite, CaCO$_3$

Mica, K$_2$O . 3Al$_2$O$_3$. 6SiO$_2$. 2H$_2$O

33   34   35   36   37   38   39   40

9   8   7

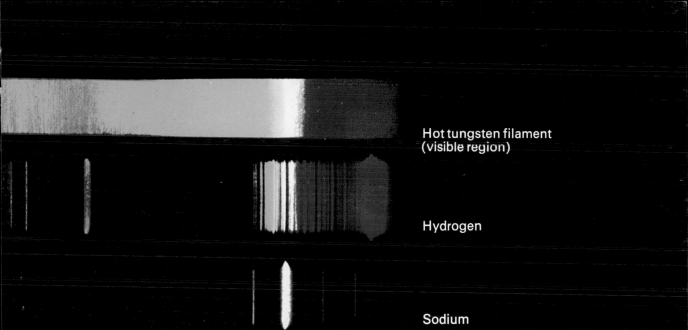

Hot tungsten filament
(visible region)

Hydrogen

Sodium

Beryl, $3BeO.Al_2O_3.6SiO_2$

Queen Victoria 1d stamp 1881
(dyed with mauve)

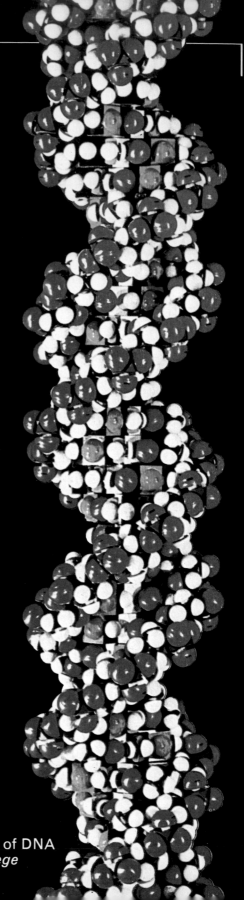

Model of the structure of the molecule of DNA
Medical Research Council, King's College

The following notes and abbreviations apply generally to Tables 5.2, 5.3, and 5.4.

1 State. The normal physical state of the material at 298 K and 1 atm (101 325 Pa) is indicated as follows: s solid; l liquid; g gas; aq aqueous solution.

2 Crystal system. The crystal system for the material at 298 K and 1 atm, or close to the melting temperature in the case of liquids and gases, is indicated as follows:

CUB	cubic	HEX	hexagonal	TET	tetragonal
TRG	trigonal (but not RBL)	RBL	rhombohedral (special case of TRG)		
ORH	orthorhombic	MCL	monoclinic	TCL	triclinic

In the following cases, the structure type is indicated instead; this implies the system.

For the cubic system
$\begin{cases} \text{BCC} & \text{body-centred cubic} \\ \text{FCC} & \text{face-centred cubic (cubic close-packed)} \\ \text{DIA} & \text{diamond structure} \end{cases}$

For the hexagonal system   HCP   hexagonal close-packed

Non-crystalline solids are indicated by AMS amorphous; POW powder; VIT vitreous.

See Table 4.8 for crystal systems and structures; and Colour Tables C.1 and C.3 for photographs of some crystals and crystal models.

3 Density ρ. Values given are measured densities at 1 atm and may in some cases be significantly less than the theoretical densities derived from X-ray measurements, if the crystal contains a high concentration of imperfections. Densities are at 298 K, except where indicated or when the stable state at 1 atm is a gas, in which case the density given is for the corresponding liquid at the boiling point for 1 atm.

4 Melting and boiling temperatures, T_m and T_b. These relate to a pressure of 1 atm unless otherwise indicated. ^{sub} sublimation. ^{dec} decomposition. $^{dhd(n)}$ loses (n molecules of) water of crystallization. ^{tr} solid state transition.

5 Thermochemical data. In addition to absolute standard molar entropies ($S^\ominus$) and standard molar Gibbs free energy changes of formation ($\Delta G_f^\ominus$), standard molar enthalpy changes are given for phase transitions (melting $\Delta H_m^\ominus$, boiling $\Delta H_b^\ominus$, solid state $\Delta H_{tr}^\ominus$), atomization ($\Delta H_{at}^\ominus$), combustion ($\Delta H_c^\ominus$), and formation ($\Delta H_f^\ominus$). With a very few exceptions (as indicated), the chosen standard pressure is always 1 atm and except for phase transitions, the chosen temperature is 298 K. For melting and boiling, the chosen temperature is that for transition at 1 atm. For formation, the elements are in physical states stable at 1 atm and at the temperature concerned (here 298 K). These definitions imply that $\Delta H_f^\ominus$ and $\Delta G_f^\ominus$ are zero for elements in their standard state.

6 Notes. Almost all chemical substances are poisonous and should be treated with respect. Those marked here are those requiring *special* precautions in laboratory or industrial practice. The following abbreviations are used.

P poisonous substance; **Ps** poison absorbed through skin; **Pv (Pg)** poisonous vapour (gas) (with maximum permissible concentration in air in parts per million); **Pc** cumulative poison (with maximum permissible body burden); **R** radioactive material; **C** corrosive material; **E** explosive substance; **F** highly flammable; **B** burns spontaneously if finely divided (pyrophoric substance); **W** violent reaction with water.

ox oxidizes in air; hyg hygroscopic; dlq deliquescent; eff efflorescent.

Colours – other than white or none – are indicated as follows.

bl blue; bk black; br brown; gn green; gr grey; or orange; rd red; yl yellow; dk dark; pa pale; pu purple; vi violet.

Solid-state transitions for elements are denoted by figures in brackets: T_{tr}/K, $\Delta H_{tr}^\ominus/J\,mol^{-1}$. C indicates Curie point.

Standard densities of gaseous elements (at 273.15 K and 1 atm) may be calculated from the formula $\rho = kM/V_m^\ominus$ where M is molar mass and $V_m^\ominus$ the ideal molar volume at 273.15 K and 1 atm (22.41 dm^3 mol^{-1}). k is a correction factor given in the notes in braces. Thus {0.099940} for $k(H_2)$.

See Table 5.1 for general notes and abbreviations. For this table, the mole applies to single atoms. Care may be needed when interpreting the results for $N \neq 1$.

Z Atomic number.
N Number of atoms per molecule in the most stable gaseous state at T_b.
St State: s solid; l liquid; g gaseous.
Cs Crystal system, see Table 5.1.
A Molar mass of element.
ρ Density (at 298 K) or density of liquid at T_b for gases.
T_m Melting temperature $\Big\{$ at 1 atm except where
T_b Boiling temperature $\Big\{$ otherwise stated.

$\Delta H_m^\ominus$ Standard molar enthalpy change of fusion at T_m.
$\Delta H_b^\ominus$ Standard molar enthalpy change of vaporization at T_b.
$S^\ominus$ Standard molar entropy at 298 K.
$\Delta H_{at}^\ominus$ Standard molar enthalpy change of atomization at 298 K from the stable state at 1 atm.
Notes See Table 5.1 for abbreviations used.

$\Big\{$ chosen standard pressure is 1 atm.

Z	Element		Z	N	St	Cs	A /g mol⁻¹	ρ /g cm⁻³	T_m /K	T_b /K	$\Delta H_m^\ominus$ /kJ mol⁻¹	$\Delta H_b^\ominus$ /kJ mol⁻¹	$S^\ominus$ /J mol⁻¹ K⁻¹	$\Delta H_{at}^\ominus$ /kJ mol⁻¹	Notes
1	Hydrogen	H		2	g	HCP	1.0	$0.07^{20\,K}$	14	20	0.06	0.45	65.3	218.0	E; {0.99940}
2	Helium	He		1	g	HCP‡	4.0	$0.15^{3\,K}$	$1^{26\,atm}$	4	0.02	0.08	126.0	—	[2.18, 0]; {0.9984}
3	Lithium	Li		1	s	BCC	6.9	0.53	454	1615	3.02	134.68	29.1	159.4	C; ox; [77.]
4	Beryllium	Be		1	s	HCP	9.0	1.85	1551	3243	12.50	294.6	9.5	324.3	P (0.001 in air)
5	Boron	B		1	s	TET	10.8	2.34	2573	2823^{sub}	22.18	538.9	5.9	562.7	
6	Carbon (graphite)	C		1‡	s	HEX	12.0	2.25*	3925–70^{sub}	5100	—	716.7^{sub}	5.7	716.7	bk
6	Carbon (diamond)ᴬ	C		1‡	s	DIA	12.0	3.51	>3823	5100	—	—	2.4	714.8	—
7	Nitrogen	N		2	g	HCP	14.0	$0.81^{77\,K}$	63	77	0.36	2.79	95.8	472.7	[36, 23]
8	Oxygen	O		2	g	CUB	16.0	$1.15^{90\,K}$	55	90	0.22	3.41	102.5	249.2	[24, 94]; [44, 743] dk bl liq
9	Fluorine	F		2	g		19.0	$1.51^{85\,K}$	53	85	2.55	3.27	158.6	79.0	Pc(0.0001); pa yl [460, 728]
10	Neon	Ne		1	g	FCC	20.2	$1.20^{27\,K}$	25	27	0.34	1.77	146.2	—	{0.99941}
11	Sodium	Na		1	s	BCC	23.0	0.97	371	1156	2.60	89.04	51.2	107.3	B; W; ox
12	Magnesium	Mg		1	s	HCP	24.3	1.74	922	1380	8.95	128.66	32.7	147.7	B
13	Aluminium	Al		1	s	FCC	27.0	2.70	933	2740	10.67	293.72	28.3	326.4	—
14	Silicon	Si		1	s	DIA	28.1	2.32–4	1683	2628	46.44	376.8	18.8	455.6	—

† Uncertain. ‡ Highly uncertain. * Variable. ^{sub} Sublimes.
ᴬ ΔH_{tr} to graphite = 1.90 kJ mol⁻¹; ΔG_{tr} to graphite = 2.90 kJ mol⁻¹.

Z	Element	N	St	Cs	A/g mol⁻¹	ρ/g cm⁻³	T_m/K	T_b/K	$\Delta H_m^{\ominus}$/kJ mol⁻¹	$\Delta H_b^{\ominus}$/kJ mol⁻¹	$S^{\ominus}$/J mol⁻¹ K⁻¹	$\Delta H_{at}^{\ominus}$/kJ mol⁻¹	Notes
15	Phosphorus (red)	P	4	s	MCL†	31.0	2.34	$863^{43\,atm}$	473^{ign}	$4.71^{143\,atm}$	30.1^{sub}	22.8	332.2	P
15	Phosphorus (white)	P	4	s	CUB	31.0	1.82	317	553	0.63	12.4	41.1	314.6	P; B
15	Phosphorus (black)^B	P	4	s	MCL	31.0	2.70*	—	—	—	—	—	354.0	P
16	Sulphur (rhombic)	S	8	s	ORH	32.1	2.07	386	—	—	—	31.8	278.8	[369, 0.38]^C
16	Sulphur (monoclinic)	S	8	s	MCL	32.1	1.96	392	718	1.41	9.62	32.6	278.5	sol in CS_2
17	Chlorine	Cl	2	g	TET	35.5	$1.56^{238\,K}$	172	238	3.20	10.20	82.5	121.7	Pv(1.0); {1 0160}; yl-gn {1.0009}
18	Argon	Ar	1	g	FCC	39.9	$1.40^{87\,K}$	84	87	1.18	6.52	154.7	—	—
19	Potassium	K	1	s	BCC	39.1	0.86	336	1033	2.32	77.53	64.2	89.2	B; W; cx
20	Calcium	Ca	1	s	FCC	40.1	1.54	1112	1757	8.66	149.95	41.4	178.2	ox; [713†, 1130]
21	Scandium	Sc	1	s	HCP	45.0	2.99	1814	3104	16.11	304.80	34.6	377.8	—
22	Titanium	Ti	1	s	HCP	47.9	4.5	1933	3560	15.48	428.86	30.6	469.9	[1155, 3975]
23	Vanadium	V	1	s	BCC	50.9	5.96	2163†	3653	17.57	458.57	28.9	514.2	P
24	Chromium	Cr	1	s	BCC	52.0	7.20	2130†	2943	13.81	348.78	23.8	396.6	[2113, 1464] [1000, 2238]; [1374, 2280]; [1410, 1799]
25	Manganese	Mn	1	s	CUB	54.9	7.20	1517	2235	14.64	219.74	32.0	280.7	[1033, 0 C]; [1183, 900]; [1673, 690]
26	Iron	Fe	1	s	BCC	55.9	7.86	1808	3023	15.36	351.04	27.3	416.3	—
27	Cobalt	Co	1	s	FCC	58.9	8.9	1768	3143	15.23	382.42	30.0	424.7	[720, 251]; [1395, 544 C]
28	Nickel	Ni	1	s	FCC	58.7	8.90	1728	3003	17.61	371.83	29.9	429.7	[680, 377 C]
29	Copper	Cu	1	s	FCC	63.5	8.92	1356	2840	13.05	304.60	33.2	338.3	yl-rd
30	Zinc	Zn	1	s	HCP^D	65.4	7.14	693	1180	7.38	115.31	41.6	130.7	—
31	Gallium	Ga	1	s	ORH	69.7	5.90^E	303	2676	5.59	256.06	40.9	277.0	—
32	Germanium	Ge	1	s	DIA	72.6	5.35	1210	3103	31.80	334.30	31.1	376.6	—
33	Arsenic (grey)	As	4	s	TRG	74.9	5.73	$1090^{28\,atm}$	886^{sub}	$27.61^{36\,atm}$	129.7^{sub}	35.1	302.5	Pc

† Uncertain. ‡ Highly uncertain. * Variable. ^{sub} Sublimes. ^{ign} Ignites.
^{B} ΔH_{tr} to red phosphorus $= -39.3$ kJ mol⁻¹. ^{C} tr to MCL.
^{D} Distorted, c/a = 1.9. ^{E} ρ(liq) = 6.09 g cm⁻³.

Z	Element		N	St	Cs	A/g mol⁻¹	ρ/g cm⁻³	T_m/K	T_b/K	ΔH_m/kJ mol⁻¹	ΔH_b/kJ mol⁻¹	$S^\ominus$/J mol⁻¹ K⁻¹	$\Delta H_{at}^\ominus$/kJ mol⁻¹	Notes
34	Selenium	Se	2	s	TRG	79.0	4.81	490	958	5.44	26.32	42.4	227.1	P; gr, rd or bk; [398, 4393]
35	Bromine	Br	2	l	ORH	79.9	3.12²⁹³ᴷ	266	332	5.27	15.00	174.9	111.9	CP; rd-br {1.0028}
36	Krypton	Kr	1	g	FCC	83.8	2.15¹²¹ᴷ	116	121	1.64	9.03	164.0	—	
37	Rubidium	Rb	1	s	BCC	85.5	1.53	312	959	2.34	69.20	76.8	80.9	B; W; ox; [243,]
38	Strontium	Sr	1	s	FCC	87.6	2.6	1042	1657	9.20	138.91	52.3	164.4	B; [486,]; [862, 837‡]
39	Yttrium	Y	1	s	HCP	88.9	4.47	1795	3611	17.15	393.30	44.4	421.3	—
40	Zirconium	Zr	1	s	HCP	91.2	6.49	2125	4650	16.74	581.58	39.0	608.8	B; [1135, 3828]
41	Niobium	Nb	1	s	BCC	92.9	8.57	2740†	5015	26.78	696.64	36.4	725.9	—
42	Molybdenum	Mo	1	s	BCC	95.9	10.2	2883	5833	27.61	594.13	28.7	658.1	—
43	Technetium	Tc	1	s	HCP	99.0	11.50ᶠ	2445	5150	23.01	577.4	33.5	678.0	R
44	Ruthenium	Ru	1	s	HCP	101.1	12.30	2583	4173	25.52	567.77	28.5	642.7	[1473, 0]; [1773, 134]
45	Rhodium	Rh	1	s	FCC	102.9	12.4	2239	4000‡	21.76	495.39	31.5	556.9	—
46	Palladium	Pd	1	s	FCC	106.4	12.02	1827	3243	16.74	393.30	37.6	378.2	—
47	Silver	Ag	1	s	FCC	107.9	10.5	1235	2485	11.30	255.06	42.6	284.6	—
48	Cadmium	Cd	1	s	HCPᴳ	112.4	8.64	594	1038	6.07	99.87	51.8	112.0	Pv(0.1)
49	Indium	In	1	s	TETᴴ	114.8	7.30	429	2353	3.26	226.35	57.8	243.3	P [476, 8]
50	Tin (white)	Sn	1	s	TET	118.7	7.28	505ᴶ	2533	7.20	290.37	51.5	302.1	[286, 2.09]ᴷ
50	Tin (grey)	Sn	1	s	DIA	118.7	5.75	505	2543	—	—	44.1	304.2	P(0.5 in air): [368,]; [690,]
51	Antimony	Sb	4	s	RBL	121.8	6.68	904	2023	19.83	67.91	45.7	262.3	P(0.01 in air)
52	Tellurium	Te	2	s	TRG	127.6	6.00	723	1263	17.49	50.63	49.7	196.7	Psᴸ; pu-bk {1.00706}
53	Iodine	I	2	s	ORH	126.9	4.93	387	457	7.89	20.85	180.7	106.8	E in H_2O
54	Xenon	Xe	1	g	FCC	131.3	3.52¹⁶⁴ᴷ	161	166	2.30	12.64	169.6	—	
55	Caesium	Cs	1	s	BCC	132.9	1.88	302*	942	2.13	65.90	85.2	76.1	P; ox; [643, 586]
56	Barium	Ba	1	s	BCC	137.3	3.51	998	1913	7.66	150.92	62.8	180.0	[110,]; [821,]; [982,]
57	Lanthanum	La	1	s	HCP	138.9	6.14	1194	3730	11.30	399.57	56.9	431.0	gr
72	Hafnium	Hf	1	s	HCP	178.5	13.31	2500†	4875	21.76	661.07	43.6	619.2	—
73	Tantalum	Ta	1	s	BCC	180.9	16.6	3269	5700‡	31.38	753.12	41.5	782.0	gr

† Uncertain. ‡ Highly uncertain. * Variable.
F Calculated, not measured. G Distorted, c/a 1.9. H Distorted FCC.
J Stable 286–434 K. K tr to white tin.
L Iodine is a skin irritant rather than a poison.

Z	Element		N	St	Cs	A g mol⁻¹	ρ g cm⁻³	T_m K	T_b K	$\Delta H_m^{\ominus}$ kJ mol⁻¹	$\Delta H_b^{\ominus}$ kJ mol⁻¹	$S^{\ominus}$ J mol⁻¹ K⁻¹	$\Delta H_{at}^{\ominus}$ kJ mol⁻¹	Notes
74	Tungsten	W	1	s	BCC	183.9	19.35	3683†	5933	35.22*	799.14	32.6	849.4	—
75	Rhenium	Re	1	s	HCP	186.2	20.53	3453	5900‡	33.05	707.10	36.9	769.9	—
76	Osmium	Os	1	s	HCP	190.2	22.48	2973	>5570	29.29	627.60	32.6	790.8	—
77	Iridium	Ir	1	s	FCC	192.2	22.42	2683	4403	26.36	563.58	35.5	665.2	—
78	Platinum	Pt	1	s	FCC	195.1	21.45	2045	4100‡	19.66	510.45	41.6	565.3	—
79	Gold	Au	1	s	FCC	197.0	18.88	1337	3353	12.36	324.43	47.4	366.1	yl
80	Mercury	Hg	1	l	RBL	200.6	13.59	234	630	2.30	59.15	76.0	61.3	Pc Ps Pv(0.1)
81	Thallium	Tl	1	s	HCP	204.4	11.85	577	1730†	4.27	162.09	64.2	182.2	Ps; [507, 377]
82	Lead	Pb	1	s	FCC	207.2	11.34	601	2013	4.77	179.41	64.8	195.0	Pc
83	Bismuth	Bi	1	s	TRG	209.0	9.80	544	1833†	10.88	151.50	56.7	207.1	—
84	Polonium	Po	2	s	RBL	210.0	9.4	527	1235	12.55	60.2	62.8	144.1	R P (7 pg in body); [370,]
85	Astatine	At	2	s	—	210.0	—	575‡	610‡	11.92	45.2	60.7	90.4	R
86	Radon	Rn	1	g	—	222.0	4.4†²¹¹ᴷ	202	211	2.90	16.40	176.1	—	R Pv
87	Francium	Fr	1	s	—	223.0	—	300‡	950‡	2.09	63.6	95.4	72.8	R
88	Radium	Ra	1	s	—	226.0	5‡	973	<1410	8.37	136.82	71.1	161.9	R Pc Pv
89	Actinium	Ac	1	s	—	227.0	10.07ᴹ	1323	3473	14.23	397.5	56.5	405.9	R
90	Thorium	Th	1	s	FCC	232.0	11.7	2023	5060†	15.65	543.92	53.4	598.3	R; [498,]; [1673, 2803‡]
91	Protactinium	Pa	1	s	TET	231.1	15.37	<1870	4300†	14.64	460.2	51.9	606.7	R
92	Uranium	U	1	s	BCC	238.0	19.05	1405	4091	15.48	422.6	50.2	535.6	R; Pc
94	Plutonium (α)	Pu	1	s	MCL	242.0	19.84	914	3505	2.09	317.1	—	—	R; Pc

(0.2 µCi in body); B: [941, 2820]; [1047, 4531‡];]: [750, 1966]

(0.5 ng in body); danger of criticality; [394, 3975]; [480, 586]; [590, 669]; [726,]

† Uncertain. ‡ Highly uncertain. * Variable.
M Calculated not measured.
References: American Society for testing materials, Gray, Wagman, Weast.

5

See Table 5.1 for general notes and abbreviations. All acids are grouped under hydrogen. Data for selected ions in an ideal gaseous or aqueous state are also included.

State s solid; l liquid; g gas; aq aqueous; coll colloidal.

Crystal system See Tables 5.1 and 4.8. Crystal structures for materials normally liquid or gaseous relate to just below T_m.

M Molar mass.

ρ Density (at 298 K) or density of liquid at just below T_b for gases unless otherwise indicated.

T_m Melting temperature ⎰at 1 atm except where

T_b Boiling temperature ⎱otherwise stated.

Compound	State	Crystal system	M / $\mathrm{g\,mol^{-1}}$	ρ / $\mathrm{g\,cm^{-3}}$	T_m / K	T_b / K
Aluminium						
AlF_3	s	HEX	84.0	2.88	1564[sub]	—
$AlCl_3$	s	HEX	133.3	2.44	463[2.5 atm]	451[sub]
$AlCl_3{\cdot}6H_2O$	s	HEX	241.4	2.40	373[dec]	—
$AlBr_3$	s	MCL[dim]	266.7	—	371	536[747]
AlI_3	s	—	407.7	3.98	464	633
Al_2O_3 (corundum)	s	HEX	102.0	3.97	2345	3253
$Al(OH)_3$	s	MCL	78.0	2.42	573[dhd]	—
$Al(NO_3)_3{\cdot}6H_2O$	s	—	321.1	—	—	—
Al_2S_3	s	HEX	150.2	2.02	1373	1773[sub]
$Al_2(SO_4)_3$	s	POW	342.1	2.71	1043[dec]	—
$Al_2(SO_4)_3{\cdot}6H_2O$	s	—	450.2	—	dec	—
$Al_2(SO_4)_3{\cdot}18H_2O^\dagger$	s	MCL	666.4	1.69[†]	360[dec]	—
Al^{3+}	g	—	—	—	—	—
Ammonium compounds (see under Nitrogen)						
Antimony						
SbH_3 (stibine)	g	—	124.8	2.26[248 K]	185	256
SbF_3	s	ORH	178.8	4.38	565	592[sub]
$SbCl_3$	s	ORH	228.1	3.14	347	556
$SbCl_5$	l	—	229.0	2.35	276	352
Sb_4O_6	s	CUB	583.0	5.2	929	1823[sub]
Sb_2S_3 (black)	s	ORH	339.7	4.64	823	1423[†]
$Sb_2(SO_4)_3$	s	—	531.7	3.63	dec	dec
Sb^{3+}	g	—	—	—	—	—
Arsenic						
AsH_3 (arsine)	g	—	77.9	2.69	157	218
AsF_3	l	—	131.9	2.67	264	336
AsF_3	g	—	131.9	—	—	—
$AsCl_3$	l	—	181.2	2.16	265	403
$AsBr_3$	s	ORH	314.6	3.54	306	494

[†] Uncertain. [‡] Highly uncertain. [sub] Sublimes. [dim] Exists as dimers. [dhd] Dehydrates. [dec] Decomposes.

See Table 5.1 for general notes and abbreviations.

$\Delta H_f^\ominus$ Standard molar enthalpy change of formation at 298 K.

$\Delta G_f^\ominus$ Standard molar Gibbs free energy change of formation at 298 K.

$S^\ominus$ Standard molar entropy at 298 K.

m_{sat} Solubility in water measured in moles per 100 g water at 298 K. A figure in brackets after the solubility gives the concentration of

⎫ Chosen standard pressure is 1 atm. ⎬ the saturated solution as moles per 100 cm³ of solution for cases where the solution density is known to be significantly different from 1 g cm⁻³. This information is not available for many compounds where it would be relevant. A superscript gives water of crystallization of solid phase when different from standard state.

Compound	State	$\dfrac{\Delta H_f^\ominus}{\text{kJ mol}^{-1}}$	$\dfrac{\Delta G_f^\ominus}{\text{kJ mol}^{-1}}$	$\dfrac{S^\ominus}{\text{J mol}^{-1}\text{K}^{-1}}$	$\dfrac{m_{sat}}{\text{mol/100 g}}$	Notes (see Table 5.1)
Aluminium						
AlF_3	s	−1504.1	−1425.1	66.4	6.71×10^{-3} ³ᴴ²ᴼ	—
$AlCl_3$	s	−704.2	−628.9	110.7	5.2×10^{-1}	C Pᴬ
$AlCl_3 \cdot 6H_2O$	s	−2691.6	—	—	3.46×10^{-1}	dlq
$AlBr_3$	s	−527.2	−488.4	163.2	dec **W**	C Pᴬ
AlI_3	s	−313.8	−300.8	159.0	dec	br plates, dlq
Al_2O_3	s	−1675.7	−1582.4	50.9	1.00×10^{-10}‡	—
$Al(OH)_3$	s	−1287.4	−1149.8	85.4	1.28×10^{-6}‡ ²⁹¹ ᴷ	—
$Al(NO_3)_3 \cdot 6H_2O$	s	−2850.5	−2203.9	467.8	—	dlq
Al_2S_3	s	−723.8	—	—	dec	yl
$Al_2(SO_4)_3$	s	−3440.8	−3100.1	239.3	9.15×10^{-2}	—
$Al_2(SO_4)_3 \cdot 6H_2O$	s	−5311.7	−4622.6	469.0	—	—
$Al_2(SO_4)_3 \cdot 18H_2O^\dagger$	s	−8878.9	−7437.5	—	1.13×10^{-1}†? ¹⁶ᴴ²ᴼ	—
Al^{3+}	g	5483.9	—	149.9	—	—

Ammonium compounds (see under Nitrogen)

Compound	State					Notes
Antimony						**All Sb compounds P**
SbH_3	g	145.1	147.7	232.7	8.92×10^{-4}	P(0.1) F
SbF_3	s	−915.5	−807.0	105.4	2.15	P
$SbCl_3$	s	−382.2	−323.7	184.0	$4.33^\dagger$	P dlq
$SbCl_5$	l	−440.2	−350.2	301.0	dec	P rd
Sb_4O_6	s	−1440.6	−1268.2	220.9	slightly soluble	P
Sb_2S_3 (black)	s	−174.9	−173.6	182.0	2.06×10^{-6}	P bk (or yl rd)
$Sb_2(SO_4)_3$	s	−2402.5	—	—	insoluble	P dlq
Sb^{3+}	g	2703.3	—	168.7	—	P
Arsenic						**All As compounds Ps Pc**
AsH_3	g	66.4	68.9	222.7	8.92×10^{-4}	Ps Pc Pv (0.1)
AsF_3	l	−956.3	−909.1	181.2	dec	Ps Pc
AsF_3	g	−920.6	−905.7	289.0	—	Ps Pc
$AsCl_3$	l	−305.0	−259.4	216.3	dec	Ps Pc
$AsBr_3$	s	−197.5	−169.0	161.1	dec	Ps Pc

† Uncertain.   ‡ Highly uncertain.   ᵈᵉᶜ Decomposes.   ᴬ Dissolve violently in cold water and decompose in hot water.

Compound	State	Crystal system	M g mol^{-1}	ρ g cm^{-3}	T_m K	T_b K
Arsenic (continued)						
As_2O_3	s	MCL	197.8	3.74	584	630[‡]
As_2O_5	s	AMS	229.8	4.32	588[dec]	—
As_2S_3 (orpiment)	s	MCL	246.0	3.43	573	980
As^{3+}	g	—	—	—	—	—
Barium						
BaH_2	s	ORH	139.4	4.21	948	1673[†]
BaF_2	s	CUB	175.3	4.89	1628	2410
$BaCl_2$	s	ORH	208.2	3.91	1236	1833
$BaCl_2 \cdot 2H_2O$	s	MCL	244.3	3.10	386[dhd]	—
$Ba(ClO_3)_2$	s	—	304.3	—	687	—
$Ba(ClO_3)_2 \cdot H_2O$	s	MCL	322.3	3.18	393[dhd 1]	dec
$Ba(ClO_4)_2$	s	HCP	336.2	3.20	778	dec
$BaBr_2$	s	ORH	297.1	4.78	1120	dec
$BaBr_2 \cdot 2H_2O$	s	MCL	333.1	3.58	348[dhd1]	393[dhd]
$Ba(BrO_3)_2$	s	MCL	393.2	—	—	—
$Ba(BrO_3)_2 \cdot H_2O$	s	MCL	411.1	3.99	533[dec]	—
BaI_2	s	ORH	391.1	5.15	1013	—
$BaI_2 \cdot 2H_2O$	s	RBL	427.2	5.15	372[dhd 1]	—
$Ba(IO_3)_2$	s	MCL	487.1	5.00	dec	—
$Ba(IO_3)_2 \cdot H_2O$	s	MCL	505.2	4.66	473[dec]	—
BaO	s	CUB	153.3	5.72	2191	2273[†]
BaO_2	s	TET	169.3	4.96	723	1073[dec]
$Ba(OH)_2$	s	ORH	171.3	4.50	681	dec
$BaCO_3$	s	ORH	197.3	4.43	1123[dec]	—
$Ba(HCO_3)_2$	aq	—	259.3	—	—	—
$Ba(NO_3)_2$	s	CUB	261.3	3.24	865	dec
BaS	s	CUB	169.4	4.25	1473	—
$BaSO_4$	s	ORH	233.4	4.50	1853	—
$BaCrO_4$	s	ORH	253.3	4.50	—	—
BaC_2O_4 (oxalate)	s	—	225.4	2.66	673[dec]	—
$BaC_2O_4 \cdot 2H_2O$	s	—	261.4	3.17	dec	—
Ba^{2+}	g	—	—	—	—	—
Beryllium						
BeF_2 (α)quartz	s	HEX	47.0	1.99	1073[sub]	—
$BeCl_2$ (α)	s	ORH	79.9	1.90	678	793
$BeCl_2 \cdot 4H_2O$	s	—	151.9	—	—	—
$BeBr_2$	s	ORH	168.8	3.47	763[sub]	793
BeO	s	HEX	25.0	3.01	2823[†]	4173[†]
$Be(OH)_2$ (α)	s	ORH	43.0	—	—	—
$Be(NO_3)_2$	s	—	133.0	1.56	333	415

[†] Uncertain.   [‡] Highly uncertain.   [dec] Decomposes.   [dhd] Dehydrates.   [dhd(n)] Dehydrates (loses n molecules of H_2O).   [sub] Sublimes.

Compound	State	$\Delta H_f^{\ominus}$ kJ mol^{-1}	$\Delta G_f^{\ominus}$ kJ mol^{-1}	$S^{\ominus}$ J mol^{-1} K^{-1}	m_{sat} mol/100 g	Notes (see Table 5.1)
Arsenic (continued)						
As_2O_3	s	-653.0	-571.0	117.0	1.04×10^{-2}	Ps Pc
As_2O_5	s	-924.9	-782.4	105.4	$2.97 \times 10^{-1\,4H_2O}$	Ps Pc dlq
As_2S_3	s	-169.0	-168.6	163.6	$2.03 \times 10^{-7\,291\,K}$	Ps Pc yl–rd
As^{3+}	g	5950.2	—	162.3	—	
Barium						All soluble Ba compounds P
BaH_2	s	-178.7	-132.2	—	dec gives H_2	P gr
BaF_2	s	-1207.1	-1156.9	96.4	$9.24 \times 10^{-4\dagger}$	P
$BaCl_2$	s	-858.6	-810.4	123.7	1.46×10^{-1}	P
$BaCl_2 \cdot 2H_2O$	s	-1460.1	-1296.5	202.9	$1.78 \times 10^{-1\dagger}$	P
$Ba(ClO_3)_2$	s	-762.7	-556.9	231.0	6.97×10^{-2}	P
$Ba(ClO_3)_2 \cdot H_2O$	s	-1069.0	—	—	$1.25 \times 10^{-1\dagger}$ (0.114)	P
$Ba(ClO_4)_2$	s	-800.0	-535.1	249.0	$8.60 \times 10^{-1\ddagger\,8H_2O,\,293\,K}$ (0.423)	P
$BaBr_2$	s	-757.3	-736.8	146.0	$3.30 \times 10^{-1\dagger}$	P
$BaBr_2 \cdot 2H_2O$	s	-1366.1	-1230.5	226.0	3.56×10^{-1}	P
$Ba(BrO_3)_2$	s	-752.7	-577.4	243.0	9.86×10^{-4}	P
$Ba(BrO_3)_2 \cdot H_2O$	s	-1054.8	-824.6	292.5	2.02×10^{-3}	P
BaI_2	s	-602.1	-609.0	167.0	5.64×10^{-1} (0.401)^{15H_2O}	P
$BaI_2 \cdot 2H_2O$	s	-1216.7	—	—	0.63	P dlq
$Ba(IO_3)_2$	s	-1027.2	-864.8	249.4	8.11×10^{-5}	P
$Ba(IO_3)_2 \cdot H_2O$	s	-1322.1	-1104.2	297.0	slightly soluble	P
BaO	s	-553.5	-525.1	70.4	2.27×10^{-2}	P
BaO_2	s	-634.3	-572.0	65.7	slightly soluble dec	P gr
$Ba(OH)_2$	s	-944.7	-855.2	99.7	$1.50 \times 10^{-2\,8H_2O}$	P
$BaCO_3$	s	-1216.3	-1137.6	112.1	9.12×10^{-6}	P
$Ba(HCO_3)_2$	aq	-1921.6	-1734.4	192.0	$2.80 \times 10^{-3\,22\,atm\,CO_2}$	P
$Ba(NO_3)_2$	s	-992.1	-796.7	213.8	3.91×10^{-2} (0.038)	P
BaS	s	-460.0	-456.0	78.2	5.29×10^{-2}	P hydrolyses in H_2O
$BaSO_4$	s	-1473.2	-1362.3	132.2	9.43×10^{-7}	P
$BaCrO_4$	s	-1428.0	-1338.8	151.9	$1.14 \times 10^{-6\dagger}$	P rd
BaC_2O_4	s	-1368.6	—	—	5.2×10^{-5}	P
$BaC_2O_4 \cdot 2H_2O$	s	-1971.1	—	—	5.20×10^{-5}	P
Ba^{2+}	g	1660.5	—	170.2	—	—
Beryllium						All Be compounds Ps
BeF_2 (α)	s	-1026.8	-979.5	53.3	$1.80^{\dagger\,after\,82d}$	Ps
$BeCl_2$ (α)	s	-490.4	-445.6	82.7	$8.96 \times 10^{-1\,4H_2O}$	Ps dlq
$BeCl_2 \cdot 4H_2O$	s	-1808.3	-1563.0	243.1	—	Ps
$BeBr_2$	s	-353.5	-354.0	112.0	soluble	Ps dlq
BeO	s	-609.6	-580.3	14.1	1.40×10^{-8}	Ps
$Be(OH)_2$ (α)	s	-902.4	-815.0	51.9	—	Ps
$Be(NO_3)_2$	s	-678.0	—	—	—	Ps dlq

† Uncertain. ‡ Highly uncertain. ^{dec} Decomposes.

Compound	State	Crystal system	$\dfrac{M}{g\,mol^{-1}}$	$\dfrac{\rho}{g\,cm^{-3}}$	$\dfrac{T_m}{K}$	$\dfrac{T_b}{K}$
Beryllium (continued)						
$Be(NO_3)_2 \cdot 3H_2O$	s	—	187.1	—	—	—
BeS	s	CUB	41.1	2.36	—	—
$BeSO_4\ (\alpha)$	s	TET	105.1	2.44	820–870^{dec}	—
$BeSO_4 \cdot 4H_2O$	s	TET	177.1	1.71	473$^{dhd(2)}$	673$^{dhd(4)}$
Be^{2+}	g	—	—	—	—	—
Bismuth						
$BiCl_3$	s	CUB	315.3	4.75	505†	720
BiOCl	s	TET	260.4	7.72	—	—
BiI_3	s	HEX	589.7	5.78$^{288\,K}$	681	ca 773
Bi_2O_3	s	BCC	496.0	8.9	1098	2163
$Bi(NO_3)_3 \cdot 5H_2O$	s	TCL	485.1	2.83	303^{dec}	—
Bi_2S_3	s	ORH	514.1	7.39	958^{dec}	—
$Bi_2(SO_4)_3$	s	—	706.1	5.08$^{288\,K}$	678^{dec}	—
Bi^{3+}	g	—	—	—	—	—
Boron						
B_2H_6 (diborane)	g	—	27.7	0.45	108	181
BF_3	g	—	67.8	2.99	129	173
BCl_3	l	—	117.2	1.35	166	286
BCl_3	g	—	117.2	—	166	286
BI_3	g	—	391.5	—	—	—
B_2O_3	s	HEX	69.6	2.46	723†	2133†
B_2O_3	s	VIT	69.6	1.81	723†	—
BN	s	HEX	24.8	2.25	3300^{sub}	—
B_2S_3	s	VIT	117.8	1.55	583	—
B^{3+}	g	—	—	—	—	—
Bromine						
Br_2	g	—	159.8	—	266	332
Br^-	g	—	—	—	—	—
Cadmium						
CdF_2	s	CUB	150.4	6.64	1373	2031
$CdCl_2$	s	HEX	183.3	4.07	841	1233
$CdCl_2 \cdot H_2O$	s	—	201.3	—	—	—
$Cd(ClO_4)_2$	aq	—	311.3	—	—	—
$Cd(ClO_4)_2 \cdot 6H_2O$	s	HEX	419.3	—	—	—
$CdBr_2$	s	HEX	272.2	5.19	840	1136
CdI_2	s	HEX	366.2	5.67	660	1069
$Cd(IO_3)_2$	s	—	462.2	6.43	dec	—
CdO	s	FCC	128.4	8.15	1773†	—
$Cd(OH)_2$	s	HEX	146.4	4.79	573^{dec}	—

† Uncertain. ‡ Highly uncertain. ^{sub} Sublimes. $^{dhd(n)}$ Dehydrates (loses n molecules of H_2O). ^{dec} Decomposes.

Compound	State	$\dfrac{\Delta H_f^{\ominus}}{\text{kJ mol}^{-1}}$	$\dfrac{\Delta G_f^{\ominus}}{\text{kJ mol}^{-1}}$	$\dfrac{S^{\ominus}}{\text{J mol}^{-1}\text{K}^{-1}}$	$\dfrac{m_{\text{sat}}}{\text{mol/100 g}}$	Notes (see Table 5.1)
Beryllium (continued)						
$Be(NO_3)_2 \cdot 3H_2O$	s	-787.8	—	—	$8.04 \times 10^{-1\,4H_2O}$	**Ps** dlq pa yl
BeS	s	234.3	-232.0	35.0	dec	**Ps**
$BeSO_4\ (\alpha)$	s	-1205.2	-1093.9	77.9	insoluble	**Ps**
$BeSO_4 \cdot 4H_2O$	s	-2423.7	-2080.7	234.0	$3.79 \times 10^{-1}\ (0.353)$	**Ps**
Be^{2+}	g	2993.3	—	136.2	—	—
Bismuth						
$BiCl_3$	s	-379.1	-315.1	177.0	dec	dlq
BiOCl	s	-366.9	-322.2	120.5	insoluble	yl (or gr-bk)
BiI_3	s	-105.0	-175.3	233.9	insoluble	rd
Bi_2O_3	s	-573.9	-493.7	151.5	insoluble	—
$Bi(NO_3)_3 \cdot 5H_2O$	s	-2002.9	—	—	dec	—
Bi_2S_3	s	-143.1	-140.6	200.4	3.6×10^{-8}	br-bk
$Bi_2(SO_4)_3$	s	-2544.3	-2583.6	—	dec	—
Bi^{3+}	g	5005.7	—	175.4	—	—
Boron						
B_2H_6	g	35.6	86.6	232.0	dec	—
BF_3	g	-1137.0	-1120.3	254.0	$4.72 \times 10^{-3\,273\,K}$	**C Pv**
BCl_3	l	-427.2	-387.4	206.3	dec	—
BCl_3	g	-403.7	-388.7	290.0	dec	—
BI_3	g	71.1	20.8	349.1	—	—
B_2O_3	s	-1272.8	-1193.7	54.0	$1.60 \times 10^{-2}\ (0.256^{376\,K})$	—
B_2O_3	s	-1254.5	-1182.4	77.8	1.58×10^{-2}	—
BN	s	-254.4	-228.4	14.8	insoluble	—
B_2S_3	s	-240.6	-229.0	57.3	dec	—
B^{3+}	g	7468.8	—	138.5	—	—
Bromine						
Br_2	g	30.9	3.1	245.4	$2.24 \times 10^{-2\,293\,K}$	**C Pv**(1) rd-br
Br^-	g	-233.9	-238.7	163.4	—	—
Cadmium						
CdF_2	s	-700.4	-647.7	77.4	2.89×10^{-2}	—
$CdCl_2$	s	-391.5	-344.0	115.3	0.76	—
$CdCl_2 \cdot H_2O$	s	-688.4	-587.1	167.8	—	—
$Cd(ClO_4)_2$	aq	-334.6	-94.8	290.8	—	—
$Cd(ClO_4)_2 \cdot 6H_2O$	s	-2052.7	—	—	—	—
$CdBr_2$	s	-316.2	-296.3	137.2	$4.13 \times 10^{-1\dagger\,4H_2O}\ (0.345)$	yl
CdI_2	s	-203.3	-201.4	161.1	0.235	gn-yl
$Cd(IO_3)_2$	s	—	-377.1	—	soluble	—
CdO	s	-258.2	-228.4	54.8	3.80×10^{-6}	br
$Cd(OH)_2$	s	-560.7	-473.6	96.0	5.14×10^{-8}	—

†Uncertain.   ‡Highly uncertain.   dec Decomposes.

Compound	State	Crystal system	M / g mol^{-1}	ρ / g cm^{-3}	T_m / K	T_b / K
Cadmium (continued)						
$Cd(CN)_2$	s	CUB	164.4	—	$>473^{dec}$	—
$Cd(NO_3)_2$	s	CUB	236.4	—	623	—
$Cd(NO_3)_2 \cdot 2H_2O$	s	—	272.4	—	—	—
$Cd(NO_3)_2 \cdot 4H_2O$	s	ORH	308.4	2.45	332.4	405
CdS	s	HEX	144.5	4.82	$2023^{100\,atm}$	1253^{sub}
$CdSO_4$	s	ORH	208.5	4.69	1273	—
$CdSO_4 \cdot 2.67H_2O$	s	MCL	256.5	3.09	$315^{dhd?\,1.67}$	—
Cd^{2+}	g	—	—	—	—	—
Caesium						
CsF	s	CUB	151.9	4.11	955	1524
$CsCl$	s	CUB	168.4	3.99	918	1563
$CsClO_3$	s	HEX	216.4	3.57	—	—
$CsClO_4$	s	CUB	232.4	3.33	523^{dec}	—
$CsBr$	s	CUB	212.8	4.44	909	1573
CsI	s	CUB	259.8	4.51	899	1553
$CsIO_4$	s	ORH	323.8	4.26	—	—
Cs_2O	s	HEX	281.8	4.25	$763^{in\,N_2}$	673^{dec}
$CsOH$	s	—	149.9	3.68	545	—
$CsHCO_3$	s	RBL	193.9	—	$448^{dhd\,\frac{1}{2}}$	—
$CsNO_3$	s	CUB	194.9	3.68	687	dec
Cs_2SO_4	s	ORH	361.9	4.24	1283	—
Cs^+	g	—	—	—	—	—
Calcium						
CaH_2	s	ORH	42.1	1.9	$1089^{in\,H_2}$	$873^{\dagger dec}$
CaF_2	s	CUB	78.1	3.18	1696	2773
$CaCl_2$	s	ORH	111.0	2.15	1055	1873
$CaCl_2 \cdot H_2O$	s	ORH	129.0	—	533	—
$CaCl_2 \cdot 2H_2O$	s	ORH	147.0	0.84	473^{dhd}	—
$CaCl_2 \cdot 4H_2O$	s	—	183.0	—	dhd	—
$CaCl_2 \cdot 6H_2O$	s	HEX	219.1	1.71	303	—
$Ca(ClO_4)_2$	s	—	239.0	2.65	543^{dec}	—
$Ca(ClO_4)_2 \cdot 4H_2O$	s	—	311.0	—	—	—
$CaBr_2$	s	ORH	199.9	3.35	1003	1083
$CaBr_2 \cdot 6H_2O$	s	HEX	307.9	2.29	311.2	422
$Ca(BrO_3)_2$	s	—	295.9	—	—	—
CaI_2	s	HEX	293.9	3.96	1057	1373
$CaI_2 \cdot 8H_2O$	s	—	438.0	—	—	—
$Ca(IO_3)_2$	s	MCL	389.9	4.52	813^{dec}	—
$Ca(IO_3)_2 \cdot H_2O$	s	—	407.9	—	—	—
$Ca(IO_3)_2 \cdot 6H_2O$	s	—	498.0	—	308^{dec}	—

$\dagger$ Uncertain. $\ddagger$ Highly uncertain. ^{dec} Decomposes. ^{sub} Sublimes. $^{dhd(n)}$ Dehydrates (loses n molecules of H_2O).

Compound	State	$\Delta H_f^\ominus$ kJ mol^{-1}	$\Delta G_f^\ominus$ kJ mol^{-1}	$S^\ominus$ J mol^{-1} K^{-1}	m_{sat} mol/100 g	Notes (see Table 5.1)
Cadmium (continued)						
$Cd(CN)_2$	s	162.3	207.9	104.2	1.03×10^{-2}	P
$Cd(NO_3)_2$	s	-456.3	-259.0	197.9	0.461	—
$Cd(NO_3)_2 \cdot 2H_2O$	s	-1055.6	-748.9	—	—	—
$Cd(NO_3)_2 \cdot 4H_2O$	s	-1649.0	-1217.1	—	0.697	hyg
CdS	s	-161.9	-156.5	64.8	1.46×10^{-11}	or-yl
$CdSO_4$	s	-933.3	-822.8	123.0	3.62×10^{-1}	—
$CdSO_4 \cdot 2.67H_2O$	s	-1729.4	-1465.3	229.6	$1.58^\dagger$	—
Cd^{2+}	g	2623.5	—	167.7	—	—
Caesium						
CsF	s	-553.5	-525.5	92.8	$3.84^{1\,H_2O}$	dlq
$CsCl$	s	-443.0	-414.5	101.2	1.13	dlq
$CsClO_3$	s	-411.7	-307.9	156.1	2.90×10^{-2}	—
$CsClO_4$	s	-443.1	-314.3	175.1	8.61×10^{-3}	—
$CsBr$	s	-405.8	-391.4	113.1	5.80×10^{-1}	dlq
CsI	s	-346.6	-340.6	123.1	3.29×10^{-1}	dlq
$CsIO_4$	s	—	-380.7	184.0	6.64×10^{-3}	—
Cs_2O	s	-345.8	-308.2	146.9	very soluble (dec)	or
$CsOH$	s	-417.2	-359.0	86.0	$2.02^{303\,K}$	dlq pa yl
$CsHCO_3$	s	-966.1	-831.8	130.0	$1.079^{258\,K}$	—
$CsNO_3$	s	-506.0	-406.6	155.2	4.70×10^{-2}	—
Cs_2SO_4	s	-1443.0	-1323.7	211.9	0.461	—
Cs^+	g	458.0	—	169.7	—	—
Calcium						
CaH_2	s	-186.2	-147.3	42.0	dec	—
CaF_2	s	-1219.6	-1167.3	68.9	2.31×10^{-5}	—
$CaCl_2$	s	-795.8	-748.1	104.6	5.36×10^{-1}	dlq
$CaCl_2 \cdot H_2O$	s	-1109.2	-1010.9	—	5.95×10^{-1}	dlq
$CaCl_2 \cdot 2H_2O$	s	-1402.9	—	—	6.65×10^{-1}	—
$CaCl_2 \cdot 4H_2O$	s	-2009.6	-1724.0	212.6	9.79×10^{-1}	—
$CaCl_2 \cdot 6H_2O$	s	-2607.9	-2205.0	284.9	7.46×10^{-1}	dlq
$Ca(ClO_4)_2$	s	-736.8	—	233.0	0.789	—
$Ca(ClO_4)_2 \cdot 4H_2O$	s	-1948.9	-1476.8	433.5	—	—
$CaBr_2$	s	-682.8	-663.6	130.0	$6.25 \times 10^{-1\dagger}$	dlq
$CaBr_2 \cdot 6H_2O$	s	$-2506 \cdot 2$	-2153.1	410.0	1.929	—
$Ca(BrO_3)_2$	s	-718.8	—	227.6	—	—
CaI_2	s	-533.5	-528.9	142.0	6.19×10^{-1}	dlq pa yl
$CaI_2 \cdot 8H_2O$	s	-2929.6	—	—	—	yl
$Ca(IO_3)_2$	s	-1002.5	-893.3	230.1	5.13×10^{-4}	—
$Ca(IO_3)_2 \cdot H_2O$	s	-1293.3	—	—	—	—
$Ca(IO_3)_2 \cdot 6H_2O$	s	$-2780 \cdot 7$	-2267.7	451.9	2.61×10^{-4}	—

† Uncertain. ‡ Highly uncertain. ^{dec} Decomposes.

Compound	State	Crystal system	M g mol^{-1}	ρ g cm^{-3}	T_m K	T_b K
Calcium (continued)						
CaO	s	CUB	56.1	3.35	2887	3123
Ca(OH)$_2$	s	HEX	74.1	2.24	853^{dhd}	dec
CaC$_2$ (carbide)	s	TCL	64.1	2.22	720	2573
CaCO$_3$ (calcite)	s	RBL	100.1	2.71	1612$^{1025\,atm}$	1172^{dec}
CaCO$_3$ (aragonite)	s	ORH	100.1	2.93	793^{tr}	1098^{dec}
Ca(NO$_3$)$_2$	s	CUB	164.1	2.50	834	—
Ca(NO$_3$)$_2$·2H$_2$O	s	—	200.1	—	—	—
Ca(NO$_3$)$_2$·3H$_2$O	s	—	218.1	—	—	—
Ca(NO$_3$)$_2$·4H$_2$O	s	MCL	236.1	1.90	316	405^{dec}
CaS	s	CUB	72.1	2.5	2673^{dec}	—
CaSO$_4$ (anhydrite)	s	ORH	136.1	2.96	1723^{tr}	—
CaSO$_4$·0.5H$_2$O	s	HEX	145.1	—	436^{dhd}	—
CaSO$_4$·2H$_2$O (gypsum)	s	MCL	172.2	2.32	401$^{dhd\,1.5}$	436^{dhd}
Ca$_3$(PO$_4$)$_2$ (β)	s	HEX	310.2	3.14	1943	—
CaCrO$_4$·2H$_2$O	s	TET	192.1	—	473^{dhd}	—
CaC$_2$O$_4$	s	—	128.1	2.20	dec	—
CaC$_2$O$_4$·H$_2$O	s	MCL	146.1	2.2	473^{dhd}	dec
CaSi$_2$	s	TET	96.2	2.5	—	—
CaSiO$_3$ (wollastonite)	s	MCL	116.2	2.5	1813	—
Ca$_2$SiO$_4$	s	MCL	172.2	3.27	2403	—
Ca^{2+}	g	—	—	—	—	—
Carbon						
CO	g	—	28.0	0.8$^{82\,K}$	74	82
CO$_2$	g	—	44.0	1.1$^{195\,K}$	217$^{5.2\,atm}$	195
HCN	l	—	27.0	0.70	259	299
C$_2$N$_2$ (cyanogen)	g	—	52.0	—	245	252
CS$_2$	l	—	76.1	1.26	162	319
C	g	—	12.0	—	—	5100
C$_2$	g	—	24.0	—	—	—
C$_3$	g	—	36.0	—	—	—

For enthalpy changes of formation of gaseous ions containing carbon, see Table 5.5, footnote on first page.

Chlorine						
Cl$_2$O	g	—	86.9	3.89$^{273\,K}$	253	277^{exp}
ClO$_2$	g	—	67.4	3.01$^{214\,K}$	214	283^{exp}
Cl$^-$	g	—	—	—	—	—

† Uncertain. ‡ Highly uncertain. ^{exp} Explosive. $^{dhd(n)}$ Dehydrates (loses n molecules of H$_2$O). ^{dec} Decomposes.

Compound	State	$\Delta H_f^\ominus$ / kJ mol^{-1}	$\Delta G_f^\ominus$ / kJ mol^{-1}	$S^\ominus$ / J mol^{-1}K^{-1}	m_{sat} / mol/100 g	Notes (see Table 5.1)
Calcium (continued)						
CaO	s	-635.1	-604.0	39.7	2.34×10^{-3}	**W**
Ca(OH)$_2$	s	-986.1	-898.6	83.4	$1.53 \times 10^{-3\,free\,of\,CO_2}$	—
CaC$_2$	s	-59.8	-64.8	69.9	dec **W**	**E** (ethyne with H$_2$O)
CaCO$_3$ (calcite)	s	-1206.9	-1128.8	92.9	1.30×10^{-5}	—
CaCO$_3$ (aragonite)	s	-1207.1	-1127.8	88.7	—	—
Ca(NO$_3$)$_2$	s	-938.4	-743.2	193.3	$6.22 \times 10^{-1\dagger}$	hyg
Ca(NO$_3$)$_2$·2H$_2$O	s	-1540.8	-1229.3	269.4	—	—
Ca(NO$_3$)$_2$·3H$_2$O	s	-1838.0	-1471.9	319.2	—	dlq
Ca(NO$_3$)$_2$·4H$_2$O	s	-2132.3	-1713.5	375.3	8.41×10^{-1}	dlq
CaS	s	-482.4	-477.4	56.5	$2.94 \times 10^{-4\,293\,K}$	—
CaSO$_4$	s	-1434.1	-1321.9	106.7	$4.66 \times 10^{-3\,dec}$	—
CaSO$_4$·0.5H$_2$O	s	-1576.7	-1436.8	130.5	$1.10 \times 10^{-3\dagger}$	plaster of Paris
CaSO$_4$·2H$_2$O	s	-2022.6	-1797.4	194.1	7.00×10^{-2}	(max m_{sat} at 313 K)
Ca$_3$(PO$_4$)$_2$ (β)	s	-4120.8	-3884.8	236.0	$6.35 \times 10^{-5}*$	—
CaCrO$_4$·2H$_2$O	s	-1379.0	-1277.4	133.9	1.07×10^{-1} (0.106)	dk yl
CaC$_2$O$_4$	s	-1360.6	—	—	5.3×10^{-6}	—
CaC$_2$O$_4$·H$_2$O	s	-1674.9	-1514.0	156.5	$4.92 \times 10^{-6\dagger}$	**P**
CaSi$_2$	s	-151.0	—	—	dec	—
CaSiO$_3$	s	-1634.9	-1549.7	81.9	$8.18 \times 10^{-5\,290\,K}$	—
Ca$_2$SiO$_4$	s	-2307.5	-2192.8	127.7	—	—
Ca^{2+}	g	1925.9	—	154.8	—	—
Carbon						
CO	g	-110.5	-137.2	197.6	$2.14 \times 10^{-5\dagger\,1\,atm\,total\,pressure}$	—
CO$_2$	g	-393.5	-394.4	213.6	$3.29 \times 10^{-3\dagger\,1\,atm\,total\,pressure}$	—
HCN	l	108.9	124.9	112.8	4.50×10^{-2}	**Pv**(20)
C$_2$N$_2$	g	307.9	296.3	242.1	$2.14 \times 10^{-2}*$	**Pv**
CS$_2$	l	89.7	65.2	151.3	2.22×10^{-3}	**F**
C	g	716.7	671.3	158.0	insoluble	—
C$_2$	g	836.8	780.4	199.3	insoluble	—
C$_3$	g	793.5	773.1	212.1	insoluble	—
Chlorine						
Cl$_2$O	g	80.3	97.9	266.1	$3.29 \times 10^{-1\,293\,K}$	dec **P**(1) yl-rd
ClO$_2$	g	102.5	120.5	256.7	$1.29 \times 10^{-1\,287\,K}$	**P**(1) yl-rd
Cl$^-$	g	-246.0	-240.0	153.1	—	—

† Uncertain. ‡ Highly uncertain. $*$ Variable. ^{dec} Decomposes.

Compound	State	Crystal system	M $\overline{\text{g mol}^{-1}}$	ρ $\overline{\text{g cm}^{-3}}$	T_m $\overline{\text{K}}$	T_b $\overline{\text{K}}$
Chromium						
CrF_3	s	HEX	109.0	3.8	>1273	1373–1473[sub]
$CrCl_3$	s	HEX	158.3	2.76	1423[†]	1573[sub]
CrO_2Cl_2	l	—	154.9	1.91	177	390
CrI_3	s	HEX	432.7	4.91	>873	—
Cr_2O_3	s	HEX	152.0	5.21	2538	4273
CrO_3	s	ORH	99.9	2.70	469	dec
$Cr_2(SO_4)_3$	s	HEX	392.2	3.01	—	—
$Cr_2(SO_4)_3 \cdot 18H_2O$	s	CUB	716.4	1.7	373[dhd 12]	—
$Cr(CO)_6$	s	ORH	220.1	1.77	383[dec]	483[exp]
Cr^{3+}	g	—	—	—		
Cobalt						
CoF_3	s	HEX	115.9	—	—	—
$CoCl_2$	s	HEX	129.8	3.36	997[in HCl]	1322
$CoCl_2 \cdot 2H_2O$	s	MCL	165.9	2.48	—	—
$CoCl_2 \cdot 6H_2O$	s	MCL	237.9	1.92	359	383[dhd]
$Co(ClO_4)_2$	aq	—	257.8	—	—	—
$Co(ClO_4)_2 \cdot 6H_2O$	s	HEX	365.9	—	1807[dec]	dec
$CoBr_2$	s	HEX	218.7	4.91	951[N2]	—
$CoBr_2 \cdot 6H_2O$	s	—	326.8	2.46	320.5[dhd 4]	403[dhd 6]
CoI_2	s	HEX	312.7	5.68	788[vac]	843[vac]
$Co(IO_3)_2$	aq	—	408.7	—	—	—
$Co(IO_3)_2 \cdot 2H_2O$	s	—	444.7	—	—	—
CoO	s	CUB	74.9	6.45	2078	—
Co_3O_4	s	CUB	240.8	6.07	1173[dec]	—
$Co(OH)_2$ (pink)	s	HEX	92.9	3.60	dec	—
$Co(NO_3)_2$	s	CUB	183.0	—	—	—
$Co(NO_3)_2 \cdot 2H_2O$	s	—	219.0	—	—	—
$Co(NO_3)_2 \cdot 3H_2O$	s	—	237.0	—	—	—
$Co(NO_3)_2 \cdot 4H_2O$	s	—	255.0	—	—	—
$Co(NO_3)_2 \cdot 6H_2O$	s	MCL	291.0	1.87	328[dhd 3]	—
$CoSO_4$	s	ORH	155.0	3.71	1008[dec]	—
$CoSO_4 \cdot 7H_2O$	s	MCL	281.1	1.95	370	693[dhd]
Co^{2+}	g	—	—	—	—	—
Copper						
CuF_2	s	MCL	101.5	4.23	1223[dec]	—
$CuF_2 \cdot 2H_2O$	s	MCL	137.6	2.93	dec	—
$CuCl$	s	CUB	99.0	4.14	703	1763
$CuCl_2$	s	MCL	134.4	3.39	893	1266[dec]

[†] Uncertain.   [‡] Highly uncertain.   [dec] Decomposes.   [dhd(n)] Dehydrates (loses n molecules of H_2O).   [sub] Sublimes.   [vac] In vacuum.

Compound	State	$\dfrac{\Delta H_f^{\ominus}}{\text{kJ mol}^{-1}}$	$\dfrac{\Delta G_f^{\ominus}}{\text{kJ mol}^{-1}}$	$\dfrac{S^{\ominus}}{\text{J mol}^{-1}\text{K}^{-1}}$	$\dfrac{m_{\text{sat}}}{\text{mol/100 g}}$	Notes (see Table 5.1)
Chromium						
CrF_3	s	-1159.0	-1088.0	93.9	insoluble	gn
$CrCl_3$	s	-556.5	-486.2	115.3	1.62	vi
CrO_2Cl_2	l	-579.5	-510.9	221.8	dec	**C P** rd fuming
CrI_3	s	-205.0	-202.5	—	—	bk
Cr_2O_3	s	-1139.7	-1058.1	81.2	1.20×10^{-9}	gn
CrO_3	s	-598.5	-501.0	—	1.69 (1.072)	dk rd **C P** (0.03)
$Cr_2(SO_4)_3$	s	-3025.0	—	—	1.63×10^{-1}	vi-rd
$Cr_2(SO_4)_3 \cdot 18H_2O$	s	-8339.5	—	—	$1.67 \times 10^{-1\,16H_2O}$	bl-vi
$Cr(CO)_6$	s	-1076.9	-975.0	—	insoluble	—
Cr^{3+}	g	5648.0	—	169.6	—	—
Cobalt						
CoF_3	g	-810.9	-707.0	94.6	decomposes to $Co(OH)_3$	br
$CoCl_2$	s	-312.5	-269.9	109.2	3.39×10^{-1}	hyg bl
$CoCl_2 \cdot 2H_2O$	s	-923.0	-764.8	188.0	—	rd-vi
$CoCl_2 \cdot 6H_2O$	s	-2115.4	-1725.5	343.0	4.33×10^{-1}	rd
$Co(ClO_4)_2$	aq	-316.7	-71.5	251.0	—	rd
$Co(ClO_4)_2 \cdot 6H_2O$	s	-2038.4	—	—	0.707	rd
$CoBr_2$	s	-220.9	-210.0	135.6	0.305	gn, dlq
$CoBr_2 \cdot 6H_2O$	s	-2020.0	—	—	—	rd-vi, dlq
CoI_2	s	-88.7	-101.3	158.2	0.508	bk, hyg
$Co(IO_3)_2$	aq	-500.8	-310.4	125.5	—	bl-vi
$Co(IO_3)_2 \cdot 2H_2O$	s	-1081.9	-795.8	267.8	—	rd
CoO	s	-237.9	-214.2	53.0	insoluble	gn-br
Co_3O_4	s	-891.0	-774.0	102.5	insoluble	bk
$Co(OH)_2$ (pink)	s	-539.7	-454.4	79.0	1.40×10^{-6}	pk
$Co(NO_3)_2$	s	-420.5	-237.0	192.0	—	—
$Co(NO_3)_2 \cdot 2H_2O$	s	-1021.7	—	—	—	—
$Co(NO_3)_2 \cdot 3H_2O$	s	-1325.9	—	—	—	—
$Co(NO_3)_2 \cdot 4H_2O$	s	-1630.5	—	—	—	—
$Co(NO_3)_2 \cdot 6H_2O$	s	-2211.2	-1655.6	—	$5.57 \times 10^{-1\dagger}$	rd
$CoSO_4$	s	-888.3	-782.4	118.0	2.34×10^{-1}	dk bl
$CoSO_4 \cdot 7H_2O$	s	-2979.9	-2473.8	406.1	$2.41 \times 10^{-1\dagger}$	pa rd
Co^{2+}	g	2841.6	—	178.8	—	—
Copper						**All Cu compounds P**
CuF_2	s	-542.7	-481.0	88.0	4.63×10^{-2}	—
$CuF_2 \cdot 2H_2O$	s	—	-981.6	—	3.42×10^{-2}	bl
$CuCl$	s	-137.2	-119.9	86.2	$6.06 \times 10^{-5\dagger\,H_2O}$	—
$CuCl_2$	s	-220.1	-175.7	108.1	2.00×10^{-3}	yl-br

† Uncertain.　‡ Highly uncertain.　^{dec} Decomposes.

Compound	State	Crystal system	$\dfrac{M}{g\,mol^{-1}}$	$\dfrac{\rho}{g\,cm^{-3}}$	$\dfrac{T_m}{K}$	$\dfrac{T_b}{K}$
Copper (continued)						
$Cu(ClO_4)_2$	aq	—	262.4	—	355.3	—
$Cu(ClO_4)_2 \cdot 6H_2O$	s	MCL	370.5	2.23	355	393[dec]
$CuBr_2$	s	MCL	223.3	4.77	771	
$CuBr_2 \cdot 4H_2O$	s	—	295.3	—	—	—
CuI	s	HEX	190.4	5.62	878	1563
$Cu(IO_3)_2$	aq	—	413.3	—	—	—
$Cu(IO_3)_2 \cdot H_2O$	s	—	431.3	—	—	—
Cu_2O	s	CUB	143.1	6.0	1508	2073[dec]
CuO	s	MCL	79.5	6.40	1599	—
$Cu(OH)_2$	s	ORH	97.6	3.37	dec	—
$Cu(NO_3)_2$	s	ORH	187.6	—	—	—
$Cu(NO_3)_2 \cdot 3H_2O$	s	—	241.6	2.32	387.5	443[−HNO₃]
$Cu(NO_3)_2 \cdot 6H_2O$	s	—	295.6	2.07	299.4[dhd 3]	—
Cu_2S	s	ORH	159.1	5.6	1373	—
CuS	s	HEX	95.6	4.6	376[tr]	493[dec]
$CuSO_4$	s	ORH	159.6	3.60	473	923[dec]
$CuSO_4 \cdot 5H_2O$	s	TCL	249.7	2.28	383[dhd 4]	423[dhd]
Cu^{2+}	g	—	—	—	—	—
Fluorine						
F_2O	g	—	54.0	1.90[40 K]	49	128
F^-	g	—	—	—	—	—
Gallium						
GaF_3	s	HEX	126.7	4.47	1073[sub in N₂]	1273[†]
$GaCl_2$	s	ORH	140.6	—	437	808
$GaCl_3$	s	—	176.0	2.47	351	474
$GaBr_3$	s	—	309.5	3.69	395	552
GaI_3	s	ORH	450.4	4.15	485	618[sub]
Ga_2O_3 (β)	s	MCL	187.4	5.88	2068	—
Ga^{3+}	g	—	—	—	—	—
Germanium						
GeF_4	g	—	148.6	2.46[236.5 K]	236[sub]	623[dec]
$GeCl_2$	s	—	143.5	—	dec	—
$GeCl_4$	l	—	214.4	1.84	224	357
$GeBr_4$	l	—	392.2	—	299.1	459.5
$GeBr_4$	g	—	392.2	—	—	—
GeO	s	—	88.6	—	983[sub]	—
GeO_2	s	HEX	104.6	4.23	1388	—

[†] Uncertain. [‡] Highly uncertain. [dec] Decomposes. [sub] Sublimes. [dhd(n)] Dehydrates (loses n molecules of H_2O). [tr] Transition.

Compound	State	$\Delta H_f^{\ominus}$ kJ mol^{-1}	$\Delta G_f^{\ominus}$ kJ mol^{-1}	$S^{\ominus}$ J mol^{-1} K^{-1}	m_{sat} mol/100 g	Notes (see Table 5.1)
Copper (continued)						**All Cu compounds P**
$Cu(ClO_4)_2$	aq	-193.9	48.3	264.4	soluble	—
$Cu(ClO_4)_2 \cdot 6H_2O$	s	-1928.4	—	—	v. soluble	lt bl, dlq
$CuBr_2$	s	-141.8	-108.7	118.0	v. soluble	bk, dlq
$CuBr_2 \cdot 4H_2O$	s	-1326.3	-1081.1	293.7	—	—
CuI	s	-67.7	-69.5	96.7	4.20×10^{-6}	bn-wh
$Cu(IO_3)_2$	aq	-377.8	-190.4	137.2	—	gn
$Cu(IO_3)_2 \cdot H_2O$	s	-692.0	-468.6	247.2	—	gn
Cu_2O	s	-168.6	-146.0	93.1	insoluble	rd
CuO	s	-157.3	-129.7	42.6	$3.00 \times 10^{-6\dagger dec}$	bk
$Cu(OH)_2$	s	-449.8	-359.4	75.0	insoluble (dec)	bl
$Cu(NO_3)_2$	s	-302.9	-118.2	193.3	—	—
$Cu(NO_3)_2 \cdot 3H_2O$	s	-1217.1	—	—	0.570	bl
$Cu(NO_3)_2 \cdot 6H_2O$	s	-2110.8	—	—	0.824	bl, dlq
Cu_2S	s	-79.0	-86.2	120.9	1.20×10^{-15}	bk
CuS	s	-53.1	-53.6	66.5	2.60×10^{-16}	bk
$CuSO_4$	s	-771.4	-661.9	109.0	—	white
$CuSO_4 \cdot 5H_2O$	s	-2279.6	-1880.1	300.4	1.39×10^{-1} (0.138)	blue vitriol
Cu^{2+}	g	3054.0	—	179.0	—	—
Fluorine						
F_2O	g	-21.7	-4.6	247.3	sl. soluble (dec)	**Pv**
F^-	g	-270.7	-266.6	145.4	—	—
Gallium						
GaF_3	s	-1163.0	-1085.3	84.0	1.58×10^{-5}	—
$GaCl_2$	s	—	—	—	dec	dlq
$GaCl_3$	s	-524.7	-454.8	142.0	v. soluble	—
$GaBr_3$	s	-386.6	-359.8	180.0	soluble	—
GaI_3	s	-238.9	-217.6	49.0	dec	yl
Ga_2O_3 (β)	s	-1089.1	-998.3	85.0	insoluble	—
Ga^{3+}	g	5816.0	—	161.6	—	—
Germanium						
GeF_4	g	—	—	302.8	dec	—
$GeCl_2$	s	—	—	—	—	—
$GeCl_4$	l	-531.8	-462.8	245.6	dec	bk
$GeBr_4$	l	-347.7	-331.4	280.7	—	gy-wh
$GeBr_4$	g	-300.0	-318.0	396.1	—	—
GeO	s	-212.1	-237.2	50.0	2.00×10^{-5}	(also an insoluble form)
GeO_2	s	-551.0	-497.1	55.3	4.51×10^{-3}	—

† Uncertain. ‡ Highly uncertain. ^{dec} Decomposes.

Compound	State	Crystal system	M / g mol^{-1}	ρ / g cm^{-3}	T_m / K	T_b / K
Germanium (continued)						
GeS	s	ORH	104.6	4.01	803	703^{sub}
GeS$_2$	s	ORH	136.7	2.94	1073	$>873^{\text{sub}}$
Ge^{4+}	g	—	—	—	—	—
Gold						
AuH	g	—	198.0	—	—	—
AuF$_3$	s	HEX	254.0	—	—	—
AuCl$_3$	s	MCL	303.3	3.9	527^{dec}	538$^{\text{sub in Cl}}$
AuCl$_3$·2H$_2$O	s	—	339.3	—	—	—
AuBr$_3$	s	—	436.7	—	371$^{-\text{Br}}$, 433	—
AuI	s	TET	323.9	8.25	393^{dec}	—
Au$_2$O$_3$	s	HEX	441.9	—	433$^{\text{dec}-\text{O}}$	523$^{\text{dec}-3\text{O}}$
Au$^+$	g	—	—	—	—	—
Hydrogen (acids)						
HF	g	—	20.0	0.99	190	293 –
HCl	g	—	36.5	1.64$^{159\,\text{K}}$	158	188 –
HBr	g	—	80.9	2.77$^{206\,\text{K}}$	185	206 –
HI	g	—	127.9	2.85$^{268\,\text{K}}$	222	238 –
HIO$_3$	s	ORH	175.9	4.63	383^{dec}	—
H$_2$O	l	—	18.0	1.00	273	373
H$_2$O	g	—	18.0	—	273	373
H$_2$O$_2$	l	—	34.0	1.44	273	323
H$_2$CO$_3$	aq	—	62.0	—	—	—
HNO$_3$	l	—	63.0	1.50	231	356
H$_2$S	g	—	34.1	1.54$^{188\,\text{K}}$	188	212
H$_2$S	aq	—	34.1	—		
H$_2$S$_2$	l	—	66.1	1.33$^{183\,\text{K}}$	183	344
H$_2$SO$_4$	l	—	98.1	1.84	283	611
H$_3$PO$_4$	s	MCL	98.0	1.83	316	486$^{\text{dhd}\frac{1}{2}}$
H$_3$BO$_3$ (boric, boracic)	s	TCL	61.8	1.44	442^{tr}	573$^{\text{dhd}\,1\frac{1}{2}}$
H$_2$O$^+$	g	—	—	—	—	—
OH$^+$	g	—	—	—	—	
OH$^-$	g	—	—	—	—	
H$_2$O$_2^+$	g	—	—	—	—	
H$_2$S$^+$	g	—	—	—	—	
Iodine						
I$_2$	g	—	253.8	—	387	457
IF	g	—	145.9	—	—	—
ICl (α)	s	MCL	162.3	3.18	300	371

[†] Uncertain. [‡] Highly uncertain. [sub] Sublimes. [dec] Decomposes. [dhd(n)] Dehydrates (loses n molecules of H$_2$O). [exp] Explosive. [tr] Transition.

Compound	State	$\Delta H_f^{\ominus}$ kJ mol^{-1}	$\Delta G_f^{\ominus}$ kJ mol^{-1}	$S^{\ominus}$ $\text{J mol}^{-1}\text{K}^{-1}$	m_{sat} mol/100 g	Notes (see Table 5.1)
Germanium (continued)						
GeS	s	−69.0	−71.5	71.0	2.29×10^{-3}	yl rd
GeS_2	s	−189.5	—	—	3.29×10^{-3}	—
Ge^{4+}	g	10412.3	—	—	—	—
Gold						
AuH	g	294.9	265.7	211.0	—	—
AuF_3	s	−363.6	−297.5	210.9	—	—
$AuCl_3$	s	−117.6	−55.2	147.3	7.01×10^{-1}	dk rd
$AuCl_3 \cdot 2H_2O$	s	−715.0	−519.0	226.0	—	—
$AuBr_3$	s	−53.3	−31.0	100.0	sl. soluble	bn
AuI	s	0.0	−0.2	119.2	v. sl. soluble	gn-yl
Au_2O_3	s	−3.3	76.0	—	insoluble	—
Au^+	g	1262.4	—	174.7	—	—
Hydrogen (acids)						
HF	g	−271.1	−273.2	173.7	$4.33 \times 10^{-2\,272\,K}$	**C Pv**(1)
HCl	g	−92.3	−95.2	186.8	5.97 (2.257)	**C Pv**(5)
HBr	g	−36.4	−53.4	198.6	2.39	**C Pv**(5)
HI	g	26.5	1.7	206.5	$5.56 \times 10^{-2\,0.13\,mm\,Hg}$	**C Pv**
HIO_3	s	−230.1	−144.3	118.0	$1.44^{289\,K}$	—
H_2O	l	−285.8	−237.2	69.9	—	—
H_2O	g	−241.8	−228.6	188.7	—	—
H_2O_2	l	−187.8	−120.4	109.6	∞	**C P E** with organic compounds and some metals
H_2CO_3	aq	−699.6	−623.2	187.4	—	—
HNO_3	l	−174.1	−80.8	266.3	∞	**C Pv**(10)
H_2S	g	−20.6	−33.6	205.7	9.80×10^{-3}	**Pg**(20)
H_2S	aq	−39.7	−27.9	121.3	—	—
H_2S_2	l	−23.1	—	—	dec	—
H_2SO_4	l	−814.0	−690.1	156.9	∞	**C E** if H_2O added
H_3PO_4	s	−1279.0	−1119.2	110.5	$6.83^{0.5\,H_2O}$	dlq
H_3BO_3	s	−1094.3	−969.0	88.8	4.37×10^{-2}	—
H_2O^+	g	979.9	—	—	—	—
OH^+	g	1328.4	—	—	—	—
OH^-	g	−140.9	—	—	—	—
$H_2O_2^+$	g	923.4	—	—	—	—
H_2S^+	g	995.0	—	—	—	—
Iodine						
I_2	g	62.4	19.4	260.6	—	—
IF	g	−95.6	−118.5	236.1	—	—
ICl (α)	s	−35.1	—	—	dec	or-rd

[†] Uncertain.   [‡] Highly uncertain.   [dec] Decomposes.

Compound	State	Crystal system	M g mol^{-1}	ρ g cm^{-3}	T_m K	T_b K
Iodine (continued)						
ICl_3	s	TCL^{dim}	233.3	3.12	374$^{16\,atm}$	350^{dec}
IBr	s	—	206.8	4.42	315	389^{dec}
I_2O_5	s	—	333.8	4.80	573$^{\dagger dec}$	—
I^-	g	—	—	8.25	393^{dec}	—
I_2^+	g	—	—	—	—	—
Iron						
FeF_2	s	TET	93.8	4.09	>1273	—
FeF_3	aq	—	112.8	3.52	>1273	—
$FeCl_2$	s	HEX	126.7	3.16	945	sub
$FeCl_2 \cdot 2H_2O$	s	MCL	162.8	2.36	—	—
$FeCl_2 \cdot 4H_2O$	s	MCL	198.8	1.93	—	—
$FeCl_3$	s	HEX	162.2	2.90	579	588^{dec}
$FeCl_3 \cdot 6H_2O$	s	—	270.3	—	310	553–558
$Fe(ClO_4)_2$	aq	—	254.7	—	—	—
$Fe(ClO_4)_2 \cdot 6H_2O$	s	HEX	362.8	—	>373^{dec}	—
$FeBr_2$	s	HEX	215.7	4.64	957$^{\dagger dec}$	—
FeI_2	s	HEX	309.7	5.31	red heat	—
FeI_3	g	—	436.6	—	—	—
FeO	s	CUB	71.8	5.7	1642	—
Fe_2O_3 (haematite)	s	TET	159.7	5.24	1838	—
Fe_3O_4 (magnetite)	s	CUB	231.5	5.18	1867	—
$Fe(OH)_2$	s	HEX	89.9	3.4	dec	—
$Fe(OH)_3$	s	MCL	106.8	—	—	—
$FeCO_3$ (siderite)	s	HEX	115.8	3.8	dec	—
$Fe(CO)_5$	l	—	195.9	1.46	252	376
FeS (α)	s	HEX	87.9	4.74	1468	dec
$FeSO_4$	s	ORH	151.9	—	—	—
$FeSO_4 \cdot 7H_2O$	s	MCL	278.0	1.90	337	363$^{dhd\,6}$ (573$^{dhd\,1}$)
$Fe_2(SO_4)_3$	s	MCL	399.9	3.10	753^{dec}	—
$Fe(NO_3)_3$	aq	—	242.0	—	—	—
Fe^{2+}	g	—	—	—	—	—
Fe^{3+}	g	—	—	—	—	—
Lead						
PbF_2	s	ORH	245.2	8.24	1128	1563
$PbCl_2$	s	ORH	278.1	5.85	774	1223
$PbCl_4$	l	—	349.0	3.18	258	378^{exp}
$PbBr_2$	s	ORH	367.0	6.66	646	1189
$Pb(BrO_3)_2$	s	—	463.0	—	—	—
PbI_2	s	HEX	461.0	6.16	675	1127

$\dagger$ Uncertain. $\ddagger$ Highly uncertain. ^{dec} Decomposes. $^{dhd(n)}$ Dehydrates (loses n molecules of H_2O). ^{sub} Sublimes. ^{exp} Explosive.

Compound	State	$\dfrac{\Delta H_f^{\ominus}}{\text{kJ mol}^{-1}}$	$\dfrac{\Delta G_f^{\ominus}}{\text{kJ mol}^{-1}}$	$\dfrac{S^{\ominus}}{\text{J mol}^{-1}\text{K}^{-1}}$	$\dfrac{m_{\text{sat}}}{\text{mol/100 g}}$	Notes (see Table 5.1)
Iodine (continued)						
ICl_3	s	-89.5	-22.3	167.4	dec	**C P** or-rd
IBr	s	-10.5	—	138.1	dec	dk gr
I_2O_5	s	-158.1	-38.0	—	$5.61 \times 10^{-1\,286\,K}$	**C P**
I^-	g	-196.6	-221.9	169.1	v. sl. soluble	gn-yl
I_2^+	g	967.3	—	—	—	—
Iron						
FeF_2	s	-686.0	-644.0	87.0	sl. soluble	—
FeF_3	aq	-1046.4	-841.0	357.3	sl. soluble	gn
$FeCl_2$	s	-341.8	-302.3	117.9	$6.36 \times 10^{-1\,4H_2O}$ (0.508)	**E** dlq yl-gr
$FeCl_2 \cdot 2H_2O$	s	-953.1	-797.5	—	—	—
$FeCl_2 \cdot 4H_2O$	s	-1549.3	-1275.7	—	0.805	—
$FeCl_3$	s	-399.5	-334.1	142.3	$1.73^{3.5H_2O}$ dec	**C** dlq bk-br
$FeCl_3 \cdot 6H_2O$	s	-2223.8	-1812.9	—	0.340	bn-yl, v dlq
$Fe(ClO_4)_2$	aq	-347.7	-96.1	226.4	—	—
$Fe(ClO_4)_2 \cdot 6H_2O$	s	-2068.6	—	—	0.270	—
$FeBr_2$	s	-249.8	-236.0	140.7	5.05×10^{-1}	gn-yl
FeI_2	s	-113.0	-128.4	77.0	soluble	gy, hyg
FeI_3	g	71.0	—	—	—	—
FeO	s	-271.9	-245.4	58.5	insoluble	bk
Fe_2O_3	s	-824.2	-742.2	87.4	insoluble	rd-br
Fe_3O_4	s	-1118.4	-1015.5	146.4	insoluble	bk (rd) magnetic
$Fe(OH)_2$	s	-569.0	-486.6	88.0	$6.70 \times 10^{-6\dagger}$	pa gr
$Fe(OH)_3$	s	-823.0	-696.6	106.7	3.40×10^{-7}	—
$FeCO_3$	s	-740.6	-666.7	92.9	$6.22 \times 10^{-4\,291\,K,\,1\,atm\,CO_2}$	—
$Fe(CO)_5$	l	-774.0	-705.4	338.1	insoluble	**P** yl
$FeS\ (\alpha)$	s	-100.0	-100.4	60.3	$5.01 \times 10^{-6\dagger\,291\,K}$	bk
$FeSO_4$	s	-928.4	-820.9	107.5	1.03×10^{-1}	—
$FeSO_4 \cdot 7H_2O$	s	-3014.6	-2510.3	409.2	1.94×10^{-1}	bl-gn
$Fe_2(SO_4)_3$	s	-2581.5	—	261.7	$2.18 \times 10^{-1}*$	hyg yl
$Fe(NO_3)_3$	aq	-674.9	—	—	—	—
Fe^{2+}	g	2752.2	—	177.2	—	—
Fe^{3+}	g	5714.9	—	173.8	—	—
Lead						**All Pb compounds Pc**
PbF_2	s	-664.0	-617.1	110.5	$2.45 \times 10^{-4\,333\,K}$	**Pc**
$PbCl_2$	s	-359.4	-314.1	136.0	3.90×10^{-3}	**Pc**
$PbCl_4$	l	-329.2	-259.0	—	dec	**E Pc**
$PbBr_2$	s	-278.7	-261.9	161.5	2.65×10^{-3}	**Pc**
$Pb(BrO_3)_2$	s	-134.0	-50.0	—	—	**Pc**
PbI_2	s	-175.5	-173.6	174.8	1.65×10^{-4}	**Pc** yl

† Uncertain.   ‡ Highly uncertain.   * Variable. ^{dec} Decomposes.

Compound	State	Crystal system	M / $g\,mol^{-1}$	ρ / $g\,cm^{-3}$	T_m / K	T_b / K
Lead (continued)						
PbO (litharge)	s	TET	223.2	9.53	1159	1745
Pb_3O_4 (minium)	s	TET	685.6	9.1	773[dec]	—
PbO_2	s	TET	239.2	9.37	563[dec]	—
$PbCO_3$ (cerussite)	s	ORH	267.2	6.6	588[dec]	—
$Pb(NO_3)_2$	s	CUB	331.2	4.53	743[dec]	—
PbS (galena)	s	FCC	239.2	7.5	1387	1553
$PbSO_4$	s	ORH	303.2	6.2	1443	—
$PbCrO_4$ (chrome yellow)	s	MCL	323.2	6.12	1117	dec
$Pb(C_2H_3O_2)_2 \cdot 3H_2O$	s	MCL	379.3	2.55	348[dhd 1]	473[dec]
$Pb(C_2H_5)_4$	l	—	323.4	1.66	137	473[dec]
Pb^{2+}	g	—	—	—	—	—
Lithium						
LiH	s	CUB	7.9	0.82	953	—
Li_3H_4	s	—	21.7	0.66	557	—
LiF	s	CUB	25.9	2.64	1118	1949
LiCl	s	CUB	42.4	2.07	878	1613
$LiClO_3$	s	—	90.4	1.12	400.6	573[dec]
$LiClO_4$	s	ORH	106.4	2.43	509	703[dec]
$LiClO_4 \cdot H_2O$	s	—	124.4	—	—	—
$LiClO_4 \cdot 3H_2O$	s	HEX	160.4	1.84	368	373[−2H₂O]
LiBr	s	CUB	86.8	3.46	823	1538
$LiBr \cdot H_2O$	s	—	104.3	—	—	—
$LiBr \cdot 2H_2O$	s	—	122.3	—	317[dhd 1]	—
$LiBrO_3$	s	ORH	134.8	—	—	—
LiI	s	CUB	133.8	4.08	722	1444[†]
$LiI \cdot H_2O$	s	CUB	151.8	—	—	—
$LiI \cdot 2H_2O$	s	HEX	169.9	—	—	—
$LiI \cdot 3H_2O$	s	HEX	187.9	3.48	346[dhd 1]	353[dhd 2]
$LiIO_3$	s	HEX	181.8	4.50	—	—
Li_2O	s	CUB	29.9	2.01	>1973	—
LiOH	s	TET	23.9	1.46	723	1197[dec]
$LiOH \cdot H_2O$	s	MCL	42.0	1.51	—	—
Li_2CO_3	s	MCL	73.9	2.11	996	1583[dec]
$LiHCO_3$	aq	—	68.0	—	—	—
$LiNO_3$	s	HEX	68.9	2.38	537	873[dec]
$LiNO_3 \cdot 3H_2O$	s	—	130.0	—	303[dhd 2½]	334[dhd 3]
Li_2SO_4	s	MCL	109.9	2.22	—	1118

[†] Uncertain.   [‡] Highly uncertain.   [dec] Decomposes.   [dhd(n)] Dehydrates (loses n molecules of H_2O).   [sub] Sublimes.

Compound	State	$\dfrac{\Delta H_f^{\ominus}}{\text{kJ mol}^{-1}}$	$\dfrac{\Delta G_f^{\ominus}}{\text{kJ mol}^{-1}}$	$\dfrac{S^{\ominus}}{\text{J mol}^{-1}\text{K}^{-1}}$	$\dfrac{m_{sat}}{\text{mol/100 g}}$	Notes (see Table 5.1)
Lead (continued)						
PbO	s	-217.3	-187.9	68.7	$1.08 \times 10^{-5\ 291\,K}$	**Pc** yl (massicot is ORH)
Pb_3O_4	s	-718.4	-601.2	211.3	insoluble	**Pc** red lead
PbO_2	s	-277.4	-217.4	68.6	insoluble	**Pc**
$PbCO_3$	s	-700.0	-626.3	131.0	4.12×10^{-7}	**Pc**
$Pb(NO_3)_2$	s	-451.9	-251.0	213.0	4.47×10^{-1} (0.263)	**Pc**
PbS	s	-100.4	-98.7	91.2	2.84×10^{-7} † depends on pH	**Pc** bk
$PbSO_4$	s	-919.9	-813.2	148.6	1.48×10^{-5}	**Pc**
$PbCrO_4$	s	-899.6	-819.6	152.7	5.26×10^{-8}	**Pc** yl
$Pb(C_2H_3O_2)_2 \cdot 3H_2O$	s	-1851.5	—	—	2.04×10^{-1} (0.175)	**Pc** lead ethanoate (acetate)
$Pb(C_2H_5)_4$	l	52.7	336.4	472.5	—	**Pc** tetraethyl lead
Pb^{2+}	g	916.8	—	175.3	—	—
Lithium						
LiH	s	-90.5	-68.4	20.0	dec	—
Li_3H_4	s	—	—		dec	
LiF	s	-616.0	-587.7	35.6	5.09×10^{-3}	—
LiCl	s	-408.6	-384.4	59.3	$2.00 (1.402)^{1H_2O}$	—
$LiClO_3$	s	-369.0	—	—	5.531	dlq
$LiClO_4$	s	-381.0	—	—	0.564	—
$LiClO_4 \cdot H_2O$	s	-697.1	-509.6	155.2	—	—
$LiClO_4 \cdot 3H_2O$	s	-1298.0	-1001.3	254.8	0.810	—
LiBr	s	-351.2	-342.0	74.3	$2.00 \times 10^{-2\ 2H_2O}$	dlq
$LiBr \cdot H_2O$	s	-662.6	-594.3	109.6	—	—
$LiBr \cdot 2H_2O$	s	-962.7	-840.6	162.3	2.012	—
$LiBrO_3$	s	-347.0	—	—	—	—
LiI	s	-270.4	-270.3	86.8	1.21	—
$LiI \cdot H_2O$	s	-590.3	-531.4	123.0	—	—
$LiI \cdot 2H_2O$	s	-890.4	-780.3	184.0	—	—
$LiI \cdot 3H_2O$	s	-1192.1	—	—	0.804	yl, hyg
$LiIO_3$	s	-503.4	—	—	0.442	hyg
Li_2O	s	-597.9	-561.2	37.6	dec in cold H_2O	—
LiOH	s	-484.9	-439.0	42.8	5.16×10^{-1}	C
$LiOH \cdot H_2O$	s	-788.0	-681.0	71.2	0.531	—
Li_2CO_3	s	-1215.9	-1132.1	90.4	1.75×10^{-2} (0.018)	—
$LiHCO_3$	aq	-969.6	-880.9	123.4	$1.74 \times 10^{-1\ 291\,K,\ 1\,atm\,CO_2}$	—
$LiNO_3$	s	-483.1	-381.2	90.0	1.23^{3H_2O}	—
$LiNO_3 \cdot 3H_2O$	s	-1374.4	-1103.7	223.4	—	—
Li_2SO_4	s	-1436.5	-1321.8	115.1	2.36×10^{-1}	—

† Uncertain. ‡ Highly uncertain. dec Decomposes.

Compound	State	Crystal system	M $\overline{\text{g mol}^{-1}}$	ρ $\overline{\text{g cm}^{-3}}$	T_m $\overline{\text{K}}$	T_b $\overline{\text{K}}$
Lithium (continued)						
$Li_2SO_4 \cdot H_2O$	s	MCL	127.9	—	1153	—
Li_3PO_4	s	ORH	115.8	2.54	1110	—
$LiAlH_4$	s	MCL	37.9	0.92	398^{dec}	—
Li^+	g	—	—	—	—	—
Magnesium						
MgF_2	s	TET	62.3	—	1534	2512
$MgCl_2$	s	HEX	95.2	2.32	987	1685
$MgCl_2 \cdot H_2O$	s	—	113.2	—	—	—
$MgCl_2 \cdot 2H_2O$	s	—	131.2	—	—	—
$MgCl_2 \cdot 4H_2O$	s	—	167.2	—	—	—
$MgCl_2 \cdot 6H_2O$	s	MCL	203.3	1.57	390^{dec}	—
$Mg(ClO_4)_2$	s	—	223.2	2.21	524^{dec}	—
$Mg(ClO_4)_2 \cdot 2H_2O$	s	—	259.2	—	—	—
$Mg(ClO_4)_2 \cdot 4H_2O$	s	—	295.2	—	—	—
$Mg(ClO_4)_2 \cdot 6H_2O$	s	ORH	331.2	1.98	458–463	—
$MgBr_2$	s	HEX	184.1	3.72	973	1503
$MgBr_2 \cdot 6H_2O$	s	HEX	292.2	2.00	445.4	—
MgI_2	s	HEX	278.1	4.43	$<910^{dec}$	—
MgO (periclase)	s	CUB	40.3	3.58	3125	3873
$Mg(OH)_2$	s	ORH	58.3	2.36	623^{dhd}	—
$MgCO_3$ (magnesite)	s	HEX	84.3	2.96	623^{dec}	1173^{-CO_2}
Mg_3N_2	s	CUB	100.9	2.71	1073^{dec}	973^{sub}
$Mg(NO_3)_2$	s	CUB	148.3	—	—	—
$Mg(NO_3)_2 \cdot 2H_2O$	s	—	184.3	2.03	402	—
$Mg(NO_3)_2 \cdot 6H_2O$	s	MCL	256.4	1.64	362	603^{dec}
MgS	s	CUB	56.4	2.84	2273^{dec}	—
$MgSO_4$	s	ORH	120.4	2.66	1397^{dec}	—
$MgSO_4 \cdot 2H_2O$	s	MCL	156.4	—	—	—
$MgSO_4 \cdot 4H_2O$	s	—	192.5	—	—	—
$MgSO_4 \cdot 6H_2O$	s	MCL	228.5	—	—	—
$MgSO_4 \cdot 7H_2O$	s	ORH	246.5	1.68	$423^{dhd\ 6}$	473^{dhd}
$Mg_3(PO_4)_2 \cdot 4H_2O$	s	MCL	335.0	1.64	—	—
Mg_2Si	s	CUB	76.7	1.94	1375	—
$MgSiO_3$	s	MCL	100.4	3.19	1830^{dec}	—
Mg_2SiO_4 (forsterite)	s	ORH	140.7	3.21	2183	—
Mg^{2+}	g	—	—	—	—	—

[†] Uncertain.　[‡] Highly uncertain.　[*] Variable.　[sub] Sublimes.　[dhd(n)] Dehydrates (loses n molecules of H_2O).　[dec] Decomposes.

Compound	State	$\Delta H_f^{\ominus}$ kJ mol^{-1}	$\Delta G_f^{\ominus}$ kJ mol^{-1}	$S^{\ominus}$ J mol^{-1} K^{-1}	m_{sat} mol/100 g	Notes (see Table 5.1)
Lithium (continued)						
Li$_2$SO$_4$·H$_2$O	s	−1735.5	−1565.7	163.6	0.273	—
Li$_3$PO$_4$	s	−2095.8	—	—	2.57×10^{-4}	—
LiAlH$_4$	s	−116.3	−44.8	78.7	dec **W**	**E** in H$_2$O
Li$^+$	g	679.6	650.0	132.9	—	
Magnesium						
MgF$_2$	s	−1123.4	−1070.3	57.2	1.22×10^{-4}	—
MgCl$_2$	s	−641·3	−591.8	89.6	5.57×10^{-1}	—
MgCl$_2$·H$_2$O	s	−966.6	−861.8	137.2	—	dlq
MgCl$_2$·2H$_2$O	s	−1279.7	−1118.1	179.9	—	dlq
MgCl$_2$·4H$_2$O	s	−1898.9	−1623.5	264.0	—	dlq
MgCl$_2$·6H$_2$O	s	−2499.0	−2115.0	366.1	5.77×10^{-1}	dlq
Mg(ClO$_4$)$_2$	s	−568.9	−432.2	213.0	4.48×10^{-1}	dlq
Mg(ClO$_4$)$_2$·2H$_2$O	s	−1218.7	—	—	—	—
Mg(ClO$_4$)$_2$·4H$_2$O	s	−1837.2	—	—	—	—
Mg(ClO$_4$)$_2$·6H$_2$O	s	−2445.5	−1863.1	520.9	v. soluble	—
MgBr$_2$	s	−524.3	−503.8	117.2	5.51×10^{-1} (0.453)	dlq
MgBr$_2$·6H$_2$O	s	−2410.0	−2056.0	397.0	1.081	hyd, fluor$^{X\text{-rays}}$
MgI$_2$	s	−364.0	−358.2	129.7	0.532	dlq
MgO	s	−601.7	−569.4	26.9	as Mg(OH)$_2$	—
Mg(OH)$_2$	s	−924.5	−833.6	63.2	2.00×10^{-5}	—
MgCO$_3$	s	−1095.8	−1012.1	65.7	$1.50 \times 10^{-4\,291\,K}$	—
Mg$_3$N$_2$	s	−460.7	−406.0	90.0	dec	gn-yl
Mg(NO$_3$)$_2$	s	−790.7	−589.5	164.0	—	—
Mg(NO$_3$)$_2$·2H$_2$O	s	−1409.2	—	—	soluble	—
Mg(NO$_3$)$_2$·6H$_2$O	s	−2613.3	−2080.7	452.0	4.90×10^{-1} (0.394)	—
MgS	s	−346.0	−341.8	50.3	dec	pa rd-br
MgSO$_4$	s	−1284.9	−1170.7	91.6	1.83×10^{-1}	—
MgSO$_4$·2H$_2$O	s	−1896.2	−1376.5	—	—	—
MgSO$_4$·4H$_2$O	s	−2496.6	−2138.9	—	—	—
MgSO$_4$·6H$_2$O	s	−3086.9	−2632.2	348.1	—	—
MgSO$_4$·7H$_2$O	s	−3388.7	−2871.9	372.0	3.60×10^{-1} (0.281)	Epsom salt
Mg$_3$(PO$_4$)$_2$·4H$_2$O	s	−4022.9	—	—	7.61×10^{-5}	bl
Mg$_2$Si	s	−77.8	−75.0	75.0	insoluble	dcc
MgSiO$_3$	s	−1549.0	−1462.1	67.7	insoluble	—
Mg$_2$SiO$_4$	s	−2174.0	−2055.2	95.1	insoluble	—
Mg^{2+}	g	2348.5	—	148.6	—	—

† Uncertain. ‡ Highly uncertain. ^{dec} Decomposes.

Compound	State	Crystal system	M $\mathrm{g\,mol^{-1}}$	ρ $\mathrm{g\,cm^{-3}}$	T_m K	T_b K
Manganese						
$MnCl_2$	s	HEX	125.8	2.98	923	1463
$MnCl_2 \cdot H_2O$	s	—	143.8	—	—	—
$MnCl_2 \cdot 2H_2O$	s	MCL	161.8	—	—	—
$MnCl_2 \cdot 4H_2O$	s	MCL	197.9	2.01	331	471[dhd]
$MnBr_2$	s	HEX	214.8	4.38	dec	—
$MnBr_2 \cdot H_2O$	s	—	232.8	—	—	—
$MnBr_2 \cdot 4H_2O$	s	—	286.8	—	—	—
MnI_2	aq	—	308.7	—	—	—
$MnI_2 \cdot 2H_2O$	s	—	344.7	—	—	—
$MnI_2 \cdot 4H_2O$	s	—	380.7	—	dec	—
MnO	s	CUB	70.9	5.46	2058	—
Mn_3O_4	s	CUB	228.8	4.86	1837	—
Mn_2O_3	s	CUB	157.9	4.50	1353[dec]	—
MnO_2 (pyrolusite)	s	TET	86.9	5.03	808[dec]	—
$Mn(OH)_2$	s	AMS	88.9	3.26	dec	—
$MnCO_3$	s	HEX	114.9	3.13	dec	—
$Mn(NO_3)_2$	s	ORH	179.0	—	—	—
$Mn(NO_3)_2 \cdot 6H_2O$	s	VIT	287.0	1.82	299[dhd]	—
MnS	s	CUB	87.0	3.99	1888[dec]	—
$MnSO_4$	s	ORH	151.0	3.25	973	1123[dec]
$MnSO_4 \cdot H_2O$ (α)	s	MCL	169.0	2.95	390[dec]	—
$MnSO_4 \cdot 5H_2O$	s	—	241.1	—	—	—
$MnSO_4 \cdot 4H_2O$	s	MCL	223.1	2.11	300[dec]	—
Mn^{2+}	g	—	—	—	—	—
Mercury						
Hg_2F_2	s	—	439.2	8.73	843	dec
Hg_2Cl_2 (calomel)	s	TET	472.1	7.15	673[sub]	—
$HgCl_2$	s	ORH	271.5	5.44	549	575
Hg_2Br_2	s	TET	561.0	7.31	618[sub]	—
$HgBr_2$	s	—	360.4	6.11	509	595
Hg_2I_2	s	TET	655.0	7.70	413[sub]	563[dec]
HgI_2 red (α)	s	TET	454.9	6.36	400[tr]	627
HgO red	s	ORH	216.6	11.1	773[dec]	—
$Hg(OH)_2$	aq	—	234.6	—	—	—
$Hg_2(NO_3)_2 \cdot 2H_2O$	s	MCL	561.2	4.79	343	—
HgS black	s	CUB	232.6	7.73	857	—
HgS red (cinnabar)	s	HEX	232.6	8.10	857[sub]	—

[†] Uncertain.   [‡] Highly uncertain.   [sub] Sublimes.   [dec] Decomposes.   [dhd] Dehydrates.   [tr] Transition.

Compound	State	$\dfrac{\Delta H_f^{\ominus}}{kJ\,mol^{-1}}$	$\dfrac{\Delta G_f^{\ominus}}{kJ\,mol^{-1}}$	$\dfrac{S^{\ominus}}{J\,mol^{-1}\,K^{-1}}$	$\dfrac{m_{sat}}{mol/100\,g}$	Notes (see Table 5.1)
Manganese						
$MnCl_2$	s	-481.3	-440.5	118.2	5.04×10^{-1}	dlq pa rd
$MnCl_2 \cdot H_2O$	s	-789.9	-696.2	174.1	—	dlq
$MnCl_2 \cdot 2H_2O$	s	-1092.0	-942.2	218.8	—	dlq
$MnCl_2 \cdot 4H_2O$	s	-1687.4	-1423.8	303.3	6.13×10^{-1} (0.518)	dlq pa rd
$MnBr_2$	s	-384.9	-365.7	138.0	0.593	rose
$MnBr_2 \cdot H_2O$	s	-705.0	—	—		rose
$MnBr_2 \cdot 4H_2O$	s	$-1590\cdot3$	-1292.4	291.6	—	rose
MnI_2	aq	-331.0	-250.6	152.7	—	pk
$MnI_2 \cdot 2H_2O$	s	-842.7	—	—		rose, dlq
$MnI_2 \cdot 4H_2O$	s	-1438.9	—	—	soluble	rose, dlq
MnO	s	-385.2	-362.9	59.7	3.60×10^{-6}	gn
Mn_3O_4	s	-1387.8	-1283.2	155.6	insoluble	bk
Mn_2O_3	s	-959.0	-881.2	110.5	insoluble	bk
MnO_2	s	-520.0	-465.2	53.1	insoluble	bk
$Mn(OH)_2$	s	-695.4	-615.0	99.2	2.2×10^{-6}	wh-pk
$MnCO_3$	s	-894.1	-816.7	85.8	$5.92 \times 10^{-6\ 278\,K}$	pa rd
$Mn(NO_3)_2$	s	-576.3	-503.3	168.6	—	—
$Mn(NO_3)_2 \cdot 6H_2O$	s	-2371.9	-1809.6	—	8.77×10^{-1}	pa rd
MnS	s	-214.2	-218.4	78.2	$6.90 \times 10^{-6\ 291\,K}$	gn or pa rd
$MnSO_4$	s	-1065.2	-957.4	112.1	3.44×10^{-1}	rd
$MnSO_4 \cdot H_2O$ (α)	s	$-1376\cdot5$	-1214.6	—	5.83×10^{-1}	pa rd
$MnSO_4 \cdot 5H_2O$	s	-2553.1	-2140.0	—	—	rose, dlq
$MnSO_4 \cdot 4H_2O$	s	-2258.1	-1908.3	—	4.25×10^{-1}	pa rd
Mn^{2+}	g	2519.2	—	173.6	—	—
Mercury						All Hg compounds Pc Hg(II) worse than Hg(I)
Hg_2F_2	s	-485.0	-435.6	160.7	dec Hg_2O	Pc
Hg_2Cl_2	s	-265.2	-210.8	192.5	3.75×10^{-6}	Pc
$HgCl_2$	s	-224.3	-178.7	146.0	2.69×10^{-2}	Pc corrosive sublimate
Hg_2Br_2	s	-206.9	-181.1	218.0	6.95×10^{-9}	Pc pa yl
$HgBr_2$	s	-170.7	-153.1	172.0	1.69×10^{-3}	Pc
Hg_2I_2	s	-121.3	-111.0	233.5	sl. soluble	Pc yl
HgI_2 red (α)	s	-105.4	-101.7	179.9	1.06×10^{-5}	Pc rd (yl above 400 K)
HgO red	s	-90.8	-58.6	70.3	2.37×10^{-5}	Pc rd, yl
$Hg(OH)_2$	aq	-355.2	-274.9	142.3	—	Pc
$Hg_2(NO_3)_2 \cdot 2H_2O$	s	-868.2	-563.6	—	dec	Pc
HgS black	s	-53.6	-47.7	88.3	$5.40 \times 10^{-7\ 291\,K}$	Pc bk
HgS red	s	-58.2	-50.6	82.4	—	Pc rd

† Uncertain. ‡ Highly uncertain. ^{dec} Decomposes.

Compound	State	Crystal system	$\dfrac{M}{\text{g mol}^{-1}}$	$\dfrac{\rho}{\text{g cm}^{-3}}$	$\dfrac{T_m}{K}$	$\dfrac{T_b}{K}$
Mercury (continued)						
Hg_2SO_4	s	MCL	497.2	7.56	dec	—
$HgSO_4$	s	ORH	296.6	6.47	dec	—
Hg^{2+}	g	—	—	—	—	—
Hg_2^{2+}	aq	—	—	—	—	—
Nickel						
NiF_2	s	TET	96.7	4.63	1273^{sub}	—
$NiCl_2$	s	HEX	129.6	3.55	1274	1246^{sub}
$NiCl_2 \cdot 2H_2O$	s	ORH	165.6	—	—	—
$NiCl_2 \cdot 4H_2O$	s	—	219.0	—	—	—
$NiCl_2 \cdot 6H_2O$	s	MCL	237.0	—	—	—
$Ni(ClO_4)_2$	aq	—	257.7	—	—	—
$NiBr_2$	s	HEX	218.5	5.10	1236	—
$NiBr_2 \cdot 3H_2O$	s	MCL	272.6	—	$573^{dhd\,3}$	—
$Ni(IO_3)_2$	s	—	408.5	5.07	—	—
NiO	s	CUB	74.7	6.67	2257	—
$Ni(OH)_2$	s	HEX	92.7	4.15*	503^{dec}	—
$Ni(CN)_2$	s	—	110.7	—	—	—
$Ni(NO_3)_2$	s	CUB	182.8	—	—	—
$Ni(NO_3)_2 \cdot 3H_2O$	s	—	236.8	—	—	—
$Ni(NO_3)_2 \cdot 6H_2O$	s	TCL	290.8	2.05	330	410
NiS	s	HEX	90.8	5.4	1070	—
$NiSO_4$	s	ORH	154.8	3.68	1121^{dec}	—
$NiSO_4 \cdot 4H_2O$	s	MCL	226.8	—	—	—
$NiSO_4 \cdot 6H_2O$ (α)	s	TET	262.8	—	—	—
$NiSO_4 \cdot 7H_2O$	s	ORH	280.9	1.95	372	$376^{dhd\,6}$
$Ni(CO)_4$	l	—	170.7	1.32	248	316
Ni^{2+}	aq	—	—	—	—	—
Nitrogen						
N_2H_4 (hydrazine)	l	—	32.0	1.01	275	387
NF_3	g	—	71.0	$1.54^{144\,K}$	67	144
NCl_3	l	—	120.4	1.65	<233	<344
N_2O (laughing gas)	g	—	44.0	$1.98^{182\,K}$	182	185
NO	g	—	30.0	$1.27^{123\,K}$	110	121
N_2O_3	g	—	76.0	$1.45^{171\,K}$	171	277^{dec}
NO_2	g	—	46.0	$1.49^{262\,K}$	262	294
N_2O_4	g	—	92.0	$1.45^{262\,K}$	262	294
N_2O_5	s	HEX	108.0	1.64	303	320^{dec}

For HCN, C_2N_2, see Carbon. For ammonia, see next page.

[†] Uncertain. [‡] Highly uncertain. * Variable. [sub] Sublimes. [dec] Decomposes. [dhd(n)] Dehydrates (loses n molecules of H_2O).

Compound	State	$\dfrac{\Delta H_f^{\ominus}}{kJ\,mol^{-1}}$	$\dfrac{\Delta G_f^{\ominus}}{kJ\,mol^{-1}}$	$\dfrac{S^{\ominus}}{J\,mol^{-1}K^{-1}}$	$\dfrac{m_{sat}}{mol/100\,g}$	Notes (see Table 5.1)
Mercury (continued)						
Hg_2SO_4	s	-743.1	-625.9	200.7	9.45×10^{-5}	**Pc** pa yl
$HgSO_4$	s	-707.5	-590.0	145.0	dec	**Pc**
Hg^{2+}	g	2890.4	—	174.9	—	—
Hg_2^{2+}	aq	172.3	153.6	84.5	—	—
Nickel						
NiF_2	s	-651.4	-604.2	73.6	$2.65 \times 10^{-2\,293K}$	gn
$NiCl_2$	s	-305.3	-259.1	97.7	5.06×10^{-1}	dlq yl
$NiCl_2 \cdot 2H_2O$	s	-922.2	-760.2	176.0	—	gn, dlq
$NiCl_2 \cdot 4H_2O$	s	-1516.7	-1235.1	243.0	—	—
$NiCl_2 \cdot 6H_2O$	s	-2103.2	-1713.5	344.3	1.07	gn dlq
$Ni(ClO_4)_2$	aq	-312.5	-62.8	235.1	—	—
$NiBr_2$	s	-212.1	-205.0	133.0	0.516	yl-br, dlq
$NiBr_2 \cdot 3H_2O$	s	-1146.4	—	—	0.731	yl-gn
$Ni(IO_3)_2$	s	-489.1	-326.4	213.0	2.69×10^{-3}	—
NiO	s	-239.7	-211.7	38.0	insoluble	gn-bk
$Ni(OH)_2$	s	-529.7	-447.3	88.0	1.00×10^{-5}	gn
$Ni(CN)_2$	s	127.6	—	94.1	insoluble	**P** yl-bn
$Ni(NO_3)_2$	s	-415.0	-238.0	192.0	—	—
$Ni(NO_3)_2 \cdot 3H_2O$	s	-1326.3	—	—	—	gn, dlq
$Ni(NO_3)_2 \cdot 6H_2O$	s	-2211.7	-1662.7	—	5.47×10^{-1}	dlq gn
NiS	s	-82.0	-79.5	53.0	4.0×10^{-6}	bk
$NiSO_4$	s	-872.9	-759.8	92.0	0.189	yl
$NiSO_4 \cdot 4H_2O$	s	-2104.1	—	—	—	gn
$NiSO_4 \cdot 6H_2O$ (α)	s	-2682.8	-2224.9	334.5	—	gn
$NiSO_4 \cdot 7H_2O$	s	-2976.3	-2462.2	378.9	$4.45 \times 10^{-1\,6H_2O}$	gn
$Ni(CO)_4$	l	-633.0	-588.3	313.4	$1.05 \times 10^{-4\,282K}$	**F**
Ni^{2+}	aq	-54.0	-45.6	-128.9	—	—
Nitrogen						
N_2H_4	l	50.6	149.2	121.2	v. soluble	**C**
NF_3	g	-124.7	-83.3	260.6	sl. soluble	**E** at 368 K yl
NCl_3	l	230.1	—	—	insoluble	—
N_2O	g	82.0	104.2	219.7	2.66×10^{-3}	—
NO	g	90.2	86.6	210.7	1.88×10^{-4}	**Pg**(25) (bl liq)
N_2O_3	g	83.7	139.4	312.2	soluble (dec)	**Pg**(25) rd-br
NO_2	g	33.2	51.3	240.0	soluble (dec)	**Pg**(25) yl-br
N_2O_4	g	9.2	97.8	304.2	soluble (dec)	**Pg**(25)
N_2O_5	s	-41.3	113.8	178.2	soluble (dec)	**Pg**(25)

For HCN, C_2N_2, see Carbon. For ammonia, see next page.

[†] Uncertain.   [‡] Highly uncertain.   [dec] Decomposes.

Compound	State	Crystal system	$\dfrac{M}{g\,mol^{-1}}$	$\dfrac{\rho}{g\,cm^{-3}}$	$\dfrac{T_m}{K}$	$\dfrac{T_b}{K}$
Ammonia						
NH_3	g	—	17.0	$0.77^{195\,K}$	195	240
NH_4F	s	HEX	37.0	1.01	sub	—
NH_4Cl (sal ammoniac)	s	CUB	53.5	1.53	613^{sub}	$793^{>1\,atm}$
NH_4ClO_4	s	ORH	117.5	1.95	dec	—
NH_4Br	s	CUB	97.9	2.43	725^{sub}	508^{vac}
NH_4I	s	CUB	144.9	2.51	824^{sub}	493^{vac}
NH_4IO_3	s	MCL	192.9	3.31	423^{dec}	—
NH_4OH	l	—	35.0	—	—	—
NH_4NO_3	s	ORH	80.0	1.72	443	483
$(NH_4)_2SO_4$	s	ORH	132.1	1.77	508^{dec}	$786^{>1\,atm}$
NH_4VO_3	s	ORH	117.0	2.33	473^{dec}	—
$NH_4Fe(SO_4)_2 \cdot 12H_2O$	s	CUB	482.2	1.71	313	503^{dhd}
NH_4^+	aq	—	—	—	—	—
Oxygen						
O_3 (ozone)	g	—	48.0	$2.14^{81\,K}$	81	161
For CO, CO$_2$, see Carbon.						
Phosphorus						
PH_3 (phosphine)	g	—	34.0	—	140	185
PH_4I	s	TET	161.9	2.86	292	353^{sub}
PF_3	g	—	88.0	$3.99^{122\,K}$	122	172
PF_5	g	—	126.0	—	190	198
PCl_3	l	—	137.3	1.57	161	349
PCl_5	s	TET	208.2	2.12	435^{sub}	440^{dec}
$POCl_3$	l	—	153.3	1.67	275	378
PBr_3	l	—	270.7	2.85	233	446
PBr_5	s	—	430.5	3.46	$<373^{dec}$	379^{dec}
$POBr_3$	s	—	286.7	2.82	329	463
P_4O_6	s	MCL	219.9	2.13	297	448
P_4O_{10}	s	HEX	283.9	2.39	$853^{>1\,atm}$	573^{sub}
P_2S_5	s	TCL^{dim}	222.3	2.03	$561^{\dagger}$	787
Plutonium						
PuF_3	s	HEX	299.0	9.32	1700	—
$PuCl_3$	s	HEX	348.4	5.70	1033	—
$PuBr_3$	s	ORH	481.7	6.69	954	—
PuI_3	s	ORH	622.7	6.92	1050	—
PuO_2	s	FCC	274.0	11.46	—	—

† Uncertain. ‡ Highly uncertain. ^{dec} Decomposes. ^{sub} Sublimes. ^{dim} Exists as dimers. ^{vac} In vacuum.

Compound	State	$\dfrac{\Delta H_f^{\ominus}}{\text{kJ mol}^{-1}}$	$\dfrac{\Delta G_f^{\ominus}}{\text{kJ mol}^{-1}}$	$\dfrac{S^{\ominus}}{\text{J mol}^{-1}\text{K}^{-1}}$	$\dfrac{m_{sat}}{\text{mol/100 g}}$	Notes (see Table 5.1)
Ammonia						
NH_3	g	−46.1	−16.5	192.3	$3.11^{291\,K}$ (1.79)	**C P** (100)
NH_4F	s	−464.0	−348.8	72.0	2.69	—
NH_4Cl	s	−314.4	−203.0	94.6	7.34×10^{-1}	—
NH_4ClO_4	s	−295.3	−88.9	186.2	9.14×10^{-2}	—
NH_4Br	s	−270.8	−175.3	113.0	7.99×10^{-1}	hyg
NH_4I	s	−201.4	−112.5	117.0	1.27	hyg
NH_4IO_3	s	−385.8	—	—	1.07×10^{-2}	—
NH_4OH	l	−361.2	−254.1	165.6	—	—
NH_4NO_3	s	−365.6	−184.0	151.1	2.68	**E** (above 484 K)
$(NH_4)_2SO_4$	s	−1180.9	−901.9	220.1	5.78×10^{-1}	
NH_4VO_3	s	−1053.1	−888.3	140.6	—	—
$NH_4Fe(SO_4)_2 \cdot 12H_2O$	s	—	—	—	4.62×10^{-1}	vi
NH_4^+	aq	−132.5	−79.4	113.4	—	—
Oxygen						
O_3	g	142.7	163.2	238.8	2.19×10^{-1}	**Pg**(0.1)

For CO, CO_2, see Carbon.

Compound	State	$\dfrac{\Delta H_f^{\ominus}}{\text{kJ mol}^{-1}}$	$\dfrac{\Delta G_f^{\ominus}}{\text{kJ mol}^{-1}}$	$\dfrac{S^{\ominus}}{\text{J mol}^{-1}\text{K}^{-1}}$	$\dfrac{m_{sat}}{\text{mol/100 g}}$	Notes (see Table 5.1)
Phosphorus						**Many P compounds P**
PH_3	g	5.4	13.4	210.1	8.88×10^{-4}	**Pg**(0.05)
PH_4I	s	−69.9	0.8	123.0	soluble	—
PF_3	g	−918.8	−897.5	273.1	dec	**Pg**
PF_5	g	−1595.8	—	281.0	dec	**Pg**
PCl_3	l	−319.7	−272.4	217.1	dec **W**	**C Pv**(1)
PCl_5	s	−443.5	—	166.5	dec **W**	**C Pv**(0.01) pa yl
$POCl_3$	l	−597.1	−520.9	222.5	dec **W**	**C Pv**
PBr_3	l	−184.5	−175.7	240.2	dec **W**	**C Pv** yl
PBr_5	s	−269.9	—	—	dec **W**	**C Pv**
$POBr_3$	s	−458.6	−430.5	—	dec **W**	**C Pv**
P_4O_6	s	−1640.1	—	—	dec **W**	**C** dlq
P_4O_{10}	s	−2984.0	−2697.8	228.9	dec **W**	**C** dlq
P_2S_5	s	251.0	—	—	insoluble	**C** yl-gr
Plutonium						**All Pu compounds R P**
PuF_3	s	−1569.0	−1494.0	—	insoluble	**R P** pu
$PuCl_3$	s	−955.6	−894.1	—	soluble	**R P** gn
$PuBr_3$	s	−785.8	−763.2	205.0	soluble	**R P** gn
PuI_3	s	−556.0	−556.1	234.0	soluble	**R P** gn
PuO_2	s	−1045.2	−986.0	76.0	—	**R P** yl-gn

† Uncertain.　‡ Highly uncertain.　dec Decomposes.

Compound	State	Crystal system	M $\mathrm{g\,mol^{-1}}$	ρ $\mathrm{g\,cm^{-3}}$	T_m K	T_b K
Potassium						
KF	s	CUB	58.1	2.48	1131	1778
KF·2H$_2$O	s	ORH	94.1	2.45	314	429
KCl (sylvite)	s	CUB	74.6	1.98	1043	1773^{sub}
KClO$_3$	s	MCL	122.5	2.32	629	673^{dec}
KClO$_4$	s	ORH	138.5	2.52	883†	673^{dec}
KBr	s	CUB	119.0	2.75	1007	1708
KBrO$_3$	s	HEX	167.0	3.27	707$^{>1\,atm}$	643^{dec}
KBrO$_4$	s	—	183.0		—	—
KI	s	CUB	166.0	3.13	954	1603
KIO$_3$	s	MCL	214.0	3.93	833	373^{dec}
KIO$_4$	s	TET	230.0	3.62	855	573^{dec}
K$_2$O	s	CUB	94.2	2.32	623^{dec}	—
KO$_2$	s	TET	71.1	2.14	653	dec
KOH	s	ORH	56.1	2.04	633	1593
KOH·2H$_2$O	s	MCL	92.1	—	—	—
K$_2$CO$_3$	s	MCL	138.2	2.43	1164	dec
KHCO$_3$	s	MCL	100.1	2.17	373^{*dec}	—
KNO$_2$	s	MCl	85.1	1.92	713	dec
KNO$_3$ (saltpetre)	s	ORH	101.1	2.11	607	673^{dec}
KCN	s	CUB	65.1	1.52	908	—
KCNS	s	ORH	97.2	1.89	446	773^{dec}
K$_2$S	s	CUB	110.3	1.81	1113	—
K$_2$SO$_4$	s	ORH	174.3	2.66	1342	1962
KHSO$_4$	s	ORH	136.2	2.32	487	dec
KH$_2$PO$_4$	s	TET	136.1	2.34	526	—
KMnO$_4$	s	ORH	158.0	2.70	<513^{dec}	—
K$_2$CrO$_4$	s	ORH	194.2	2.73	1241	—
K$_2$Cr$_2$O$_7$	s	TCL	294.2	2.68	671	773^{dec}
KAl(SO$_4$)$_2$	s	—	258.4	—	—	—
KAl(SO$_4$)$_2$·12H$_2$O	s	CUB	474.4	1.76	366	473^{dhd}
KCr(SO$_4$)$_2$·12H$_2$O	s	CUB	499.4	1.83	362	673^{dhd}
K$_3$Fe(CN)$_6$	s	MCL	329.3	1.85	dec	—
K$_4$Fe(CN)$_6$	s	—	368.3	—	dec	—
K$_4$Fe(CN)$_6$·3H$_2$O	s	MCL	422.4	1.85	343^{dhd}	dec
K$^+$	g	—	—	—	—	—
Rubidium						
RbH	s	—	86.5	2.6	300^{dec}	—
RbF	s	CUB	104.5	3.56	1068	1683
RbCl	s	CUB	120.9	2.80	991	1663

† Uncertain. ‡ Highly uncertain. * Variable. ^{dec} Decomposes. ^{sub} Sublimes. $^{dhd(n)}$ Dehydrates (loses n molecules of H$_2$O).

Compound	State	$\Delta H_f^{\ominus}$ kJ mol^{-1}	$\Delta G_f^{\ominus}$ kJ mol^{-1}	$S^{\ominus}$ J mol^{-1} K^{-1}	m_{sat} mol/100 g	Notes (see Table 5.1)
Potassium						
KF	s	-567.3	-537.8	66.6	1.75 (1.303)	—
KF·2H$_2$O	s	-1163.6	-1021.6	155.2	3.71	—
KCl	s	-436.7	-409.2	82.6	4.81×10^{-1} (0.417)	dlq
KClO$_3$	s	-397.7	-296.3	143.1	7.00×10^{-2} (0.068)	—
KClO$_4$	s	-432.8	-303.2	151.0	1.29×10^{-2}	—
KBr	s	-393.8	-380.7	95.9	5.70×10^{-1} (0.468)	slightly hyg
KBrO$_3$	s	-360.2	-271.2	149.2	4.88×10^{-2}	—
KBrO$_4$	s	-287.9	-174.5	170.1	—	—
KI	s	-327.9	-324.9	106.3	8.92×10^{-1} (0.619)	—
KIO$_3$	s	-501.4	-418.4	151.5	4.29×10^{-2}	—
KIO$_4$	s	-467.2	-361.4	176.0	2.23×10^{-3}	—
K$_2$O	s	-361.4	—	—	as KOH	hyg
KO$_2$	s	-284.9	-239.5	116.7	2.39×10^{-3dec}	yl
KOH	s	-424.8	-379.1	78.9	1.71	**C** dlq
KOH·2H$_2$O	s	-1051.0	-887.4	151.0	2.12	—
K$_2$CO$_3$	s	-1151.0	-1063.6	155.5	8.11×10^{-1} (0.596)	dlq
KHCO$_3$	s	-963.2	-863.6	115.5	3.62×10^{-1} (0.316)	—
KNO$_2$	s	-369.8	-306.6	152.1	$3.60^{293 K}$	pa yl
KNO$_3$	s	-494.6	-394.9	133.1	3.75×10^{-1}	—
KCN	s	-113.0	-101.9	128.5	1.10	**P** liberates HCN dlq
KCNS	s	-200.2	-178.3	124.3	2.46	dlq
K$_2$S	s	-380.7	-364.0	104.6	soluble	dlq yl-br
K$_2$SO$_4$	s	-1437.8	-1321.4	175.6	6.91×10^{-2}	—
KHSO$_4$	s	-1160.6	-1031.4	138.1	3.78×10^{-1}	**C** dlq
KH$_2$PO$_4$	s	-1568.3	-1415.9	134.9	1.09×10^{-1}	dlq
KMnO$_4$	s	-837.2	-737.6	171.7	4.83×10^{-2}	purple
K$_2$CrO$_4$	s	-1403.7	-1295.8	200.1	3.35×10^{-1}	yl
K$_2$Cr$_2$O$_7$	s	-2061.4	-1882.0	291.2	5.10×10^{-2}	rd
KAl(SO$_4$)$_2$	s	-2470.2	-2240.1	204.6	—	—
KAl(SO$_4$)$_2$·12H$_2$O	s	-6061.8	-5141.7	687.4	3.02×10^{-2}	—
KCr(SO$_4$)$_2$·12H$_2$O	s	-5777.3	—	—	4.41×10^{-2}	dk rd
K$_3$Fe(CN)$_6$	s	-249.8	-129.7	426.1	1.48×10^{-1} (0.118)	rd
K$_4$Fe(CN)$_6$	s	-594.1	-453.1	418.8	7.38×10^{-2}	—
K$_4$Fe(CN)$_6$·3H$_2$O	s	-1466.5	-1169.0	593.7	8.57×10^{-2} (0.076)	pa yl
K$^+$	g	514.3	481.2	154.4	—	
Rubidium						
RbH	s	-52.3	-32.2	—	—	—
RbF	s	-557.7	-523.4	82.1	$2.88^{291 K}$	—
RbCl	s	-435.3	-407.8	95.9	7.81×10^{-1} (0.597)	—

† Uncertain. ‡ Highly uncertain. ^{dec} Decomposes.

Compound	State	Crystal system	M $\mathrm{g\,mol^{-1}}$	ρ $\mathrm{g\,cm^{-3}}$	T_m K	T_b K
Rubidium (continued)						
$RbClO_3$	s	HEX	168.9	3.19	—	—
$RbClO_4$	s	ORH	184.9	2.80	fus	dec
$RbBr$	s	CUB	165.4	3.35	966	1613
$RbBrO_3$	s	CUB	213.4	3.68	703	—
RbI	s	CUB	212.4	3.55	920	1573
$RbIO_3$	s	MCL	260.4	4.33	dec	—
$RbOH$	s	ORH	102.5	3.20	574	—
$RbOH{\cdot}H_2O$	s	—	120.5	—	574	—
$RbOH{\cdot}2H_2O$	s	—	138.5	—	—	—
Rb_2CO_3	s	—	230.9	—	$1108^{>1\,atm}$	1013^{dec}
$RbHCO_3$	s	—	146.5	—	448^{dec}	—
$RbNO_3$	s	HEX	147.5	3.11	583	—
Rb_2S	s	CUB	203.0	2.91	803^{dec}	—
Rb_2SO_4	s	ORH	267.0	3.61	1333	1973
$RbHSO_4$	s	MCL	182.5	2.89	< red heat	—
Rb^+	g	—	—	—	—	—
Scandium						
ScF_3	s	HEX	101.9	—	—	—
$ScCl_3$	s	—	151.3	2.39	1212	1073^{sub}
Sc_2O_3	s	—	137.9	3.86	—	—
Sc^{3+}	g	—	—	—	—	—
Silicon						
SiH_4 (silane)	g	—	32.1	$0.68^{188\,K}$	88	161
SiF_4	g	—	104.1	$1.66^{178\,K}$	183	187
$SiCl_4$	l	—	169.9	1.48	203	331
$SiCl_4$	g	—	169.9	$7.59^{203\,K}$	—	—
$SiBr_4$	l	—	347.7	2.77	278.4	427
$SiBr_4$	g	—	347.7	—	—	—
SiO	g	—	44.1	2.13	>1975	2153
SiO_2 (quartz)	s	HEX	60.1	2.65	1883	2503
SiO_2 (cristobalite)	s	CUB	60.1	2.32	1996	2503
SiO_2 (tridymite)	s	MCL	60.1	2.26	1976	2503
SiC (carborundum)	s	HEX	40.1	3.22	$2973^{sub\,dec}$	—
SiS_2	s	ORH	92.2	2.02	1363^{sub}	—
Si^{4+}	g	—	—	—	—	—
Silver						
AgF	s	CUB	126.9	5.85	708	1432
$AgF{\cdot}2H_2O$	s	—	162.9	—	—	—
$AgF{\cdot}4H_2O$	s	—	198.9	—	—	—
$AgCl$	s	CUB	143.3	5.56	728	1823

[†] Uncertain. [‡] Highly uncertain. [dec] Decomposes. [sub] Sublimes.

Compound	State	$\Delta H_f^{\ominus}$ kJ mol^{-1}	$\Delta G_f^{\ominus}$ kJ mol^{-1}	$S^{\ominus}$ J mol^{-1} K^{-1}	m_{sat} mol/100 g	Notes (see Table 5.1)
Rubidium (continued)						
RbClO$_3$	s	-402.9	-300.4	151.9	2.96×10^{-2}	—
RbClO$_4$	s	-437.2	-307.7	164.0	2.70×10^{-3}	—
RbBr	s	-394.6	-381.8	110.0	7.01×10^{-1}	—
RbBrO$_3$	s	-367.3	-278.1	161.1	1.37×10^{-2}	—
RbI	s	-333.8	-328.9	118.4	7.70×10^{-1}	—
RbIO$_3$	s	—	-426.3	—	8.06×10^{-3}	—
RbOH	s	-418.2	—	84.1	$1.69^{303\,K}$	dlq gr
RbOH·H$_2$O	s	-748.9	—	—	—	—
RbOH·2H$_2$O	s	-1053.2	—	—	—	—
Rb$_2$CO$_3$	s	-1136.0	-1051.0	181.4	1.95	dlq
RbHCO$_3$	s	-963.2	-863.6	121.3	$7.88 \times 10^{-1\,293\,K}$	—
RbNO$_3$	s	-495.1	-395.8	147.3	4.45×10^{-1} (0.340)	dlq
Rb$_2$S	s	-360.7	-339.0	134.0	v. soluble	—
Rb$_2$SO$_4$	s	-1435.6	-1317.0	197.4	1.90×10^{-1} (0.168)	—
RbHSO$_4$	s	-1158.9	-1030.1	—	soluble	—
Rb$^+$	g	490.1	—	164.2	—	—
Scandium						
ScF$_3$	s	-1629.2	-1555.6	92.0	—	—
ScCl$_3$	s	-925.1	-858.0	127.2	v. soluble	—
Sc$_2$O$_3$	s	-1908.8	-1819.4	77.0	insoluble	—
Sc^{3+}	g	4627.0	—	156.3	—	—
Silicon						
SiH$_4$	g	34.3	56.9	204.5	insoluble	—
SiF$_4$	g	-1614.9	-1572.7	282.4	dec	—
SiCl$_4$	l	-687.0	-619.9	239.7	dec **W**	**C Pv**
SiCl$_4$	g	-657.0	-617.0	330.6	dec	—
SiBr$_4$	l	-457.3	-443.9	277.8	dec	—
SiBr$_4$	g	-415.4	-431.8	377.8	—	—
SiO	g	-99.6	-126.3	211.5	—	—
SiO$_2$ (quartz)	s	-910.9	-856.7	41.8		⎱ **Dangerous**
SiO$_2$ (cristobalite)	s	-909.5	-855.9	42.7	2.00×10^{-4}	**as dusts**
SiO$_2$ (tridymite)	s	-909.1	-855.3	43.5		⎰
SiC	s	-62.8	-60.2	16.5	insoluble	gr-bk
SiS$_2$	s	-207.1	-175.3	66.9	dec	—
Si^{4+}	g	10428.5	—	229.8	—	—
Silver						
AgF	s	-204.6	-186.6	80.1	1.42	dlq yl
AgF·2H$_2$O	s	-800.8	-671.1	174.9	—	—
AgF·4H$_2$O	s	-1388.3	-1147.3	268.0	—	—
AgCl	s	-127.1	-109.8	96.2	1.35×10^{-6}	—

[†] Uncertain. [‡] Highly uncertain. [dec] Decomposes.

Compound	State	Crystal system	M $\mathrm{g\,mol^{-1}}$	ρ $\mathrm{g\,cm^{-3}}$	T_{m} K	T_{b} K
Silver (continued)						
$AgClO_3$	s	TET	191.3	4.43	503	543[dec]
$AgClO_4$	s	CUB	207.3	2.81	759[dec]	—
$AgBr$	s	CUB	187.8	6.47	705	1573[dec]
$AgBrO_3$	s	TET	235.8	5.21	dec	—
AgI	s	HEX	234.8	5.68	831	1779
Ag_2O	s	CUB	231.7	7.14	503[dec]	—
Ag_2CO_3	s	MCL	275.7	6.08	491[dec]	—
$AgNO_3$	s	ORH	169.9	4.35	485	717[dec]
$AgCN$	s	HEX	133.8	3.95	593[dec]	—
Ag_2S	s	BCC	247.8	7.33	448[tr]	dec
Ag_2S	s	CUB	247.8	7.32	1098	dec
Ag_2SO_4	s	ORH	311.8	5.45	925	1358[dec]
Ag_2CrO_4	s	MCL	331.7	5.63	—	—
Ag^+	g	—	—	—	—	—
Sodium						
NaH	s	FCC	24.0	0.92	1073[dec]	—
NaF	s	CUB	42.0	2.56	1266	1968
$NaCl$	s	CUB	58.4	2.17	1074	1686
$NaClO_3$	s	CUB	106.4	2.49	521–534	dec
$NaClO_4$	s	CUB	122.4	—	755[dec]	dec
$NaBr$	s	CUB	102.9	3.20	1020	1663
$NaBr\cdot2H_2O$	s	MCL	138.9	2.18	324[dhd 2]	—
$NaBrO_3$	s	CUB	150.9	3.34	654	—
NaI	s	CUB	149.9	3.67	934	1577
$NaIO_3$	s	ORH	197.9	4.28	573[dec]	—
$NaIO_3\cdot H_2O$	s	—	215.9	—	—	—
$NaIO_3\cdot5H_2O$	s	—	287.9	—	—	—
Na_2O	s	CUB	62.0	2.27	1548[sub]	—
Na_2O_2	s	TET	78.0	2.81	733[dec]	930[dec]
$NaOH$	s	ORH	40.0	2.13	592	1663
$NaOH\cdot H_2O$	s	ORH	58.0	—	337	—
Na_2CO_3	s	MCL	106.0	2.53	1124	dec
$Na_2CO_3\cdot10H_2O$	s	MCL	286.1	1.44	306–8	306[dhd]
$NaHCO_3$	s	MCL	84.0	2.16	543[dec]	—
$NaNO_2$	s	ORH	69.0	2.17	544	593[dec]
$NaNO_3$ (nitre)	s	HEX	85.0	2.26	580	653[dec]
$NaCN$	s	CUB	49.0	—	837	1769
Na_2S	s	CUB	78.0	1.86	1453	—

[†] Uncertain.   [‡] Highly uncertain.   [dec] Decomposes.   [sub] Sublimes.   [tr] Transition.   [dhd(n)] Dehydrates (loses n molecules of H_2O).

Compound	State	$\Delta H_f^\ominus$ kJ mol^{-1}	$\Delta G_f^\ominus$ kJ mol^{-1}	$S^\ominus$ J mol^{-1} K^{-1}	m_{sat} mol/100 g	Notes (see Table 5.1)
Silver (continued)						
$AgClO_3$	s	-25.5	61.7	149.4	5.22×10^{-2}	—
$AgClO_4$	s	-31.1	77.0		2.69	
$AgBr$	s	-100.4	-96.9	107.1	7.19×10^{-8}	pa yl
$AgBrO_3$	s	-27.2	54.4	152.7	8.31×10^{-4}	—
AgI	s	-61.8	-66.2	115.5	1.11×10^{-8}	yl (α), or (β) ($T_{tr} = 419$ K)
Ag_2O	s	-31.0	-11.2	121.3	2.00×10^{-5}	br-bk
Ag_2CO_3	s	-505.8	-436.8	167.4	1.20×10^{-5}	yl
$AgNO_3$	s	-124.4	-33.5	140.9	1.42	C
$AgCN$	s	146.0	156.9	107.2	5.25×10^{-5}	—
Ag_2S	s	-32.6	-40.7	144.0	2.48×10^{-16}	gr-bk
Ag_2S	s	-29.4	-39.5	150.6	2.48×10^{-16}	—
Ag_2SO_4	s	-715.9	-618.5	200.4	1.83×10^{-3}	—
Ag_2CrO_4	s	-712.1	-621.7	216.7	9.92×10^{-5}	rd
Ag^+	g	1019.2	—	167.2	—	—
Sodium						
NaH	s	-56.3	-33.5	40.0	dec	—
NaF	s	-573.6	-543.5	51.5	9.87×10^{-2} (0.098)	P
$NaCl$	s	-411.2	-384.2	72.1	6.15×10^{-1} (0.542)	hyg
$NaClO_3$	s	-365.8	-262.3	123.4	0.742	—
$NaClO_4$	s	-383.3	-254.9	142.3	soluble	—
$NaBr$	s	-361.1	-349.0	86.8	9.19×10^{-1} (0.728)	hyg
$NaBr \cdot 2H_2O$	s	-951.9	-828.4	179.1	0.572	—
$NaBrO_3$	s	-334.1	-242.8	128.9	0.182	—
NaI	s	-287.8	-286.1	98.5	1.23 (0.829)	—
$NaIO_3$	s	-481.8	—	135.1	6.75×10^{-2}	—
$NaIO_3 \cdot H_2O$	s	-779.5	-634.1	162.3	—	—
$NaIO_3 \cdot 5H_2O$	s	-1952.3	—	—	—	—
Na_2O	s	-414.2	-375.5	75.1	dec	C dlq yl
Na_2O_2	s	-510.9	-447.7	95.0	dec	pa yl
$NaOH$	s	-425.6	-379.5	64.5	1.05	C
$NaOH \cdot H_2O$	s	-734.5	-629.4	99.5	1.97	—
Na_2CO_3	s	-1130.7	-1044.5	135.0	6.60×10^{-2}	hyg
$Na_2CO_3 \cdot 10H_2O$	s	-4081.3	-3428.2	564.0	1.03×10^{-1} (0.099)	washing soda
$NaHCO_3$	s	-950.8	-851.0	101.7	1.22×10^{-1}	baking soda
$NaNO_2$	s	-358.7	-284.6	103.8	1.23 (0.898)	pa yl
$NaNO_3$	s	-467.9	-367.1	116.5	1.08	—
$NaCN$	s	-87.5	-76.4	115.6	1.29	P liberates HCN dlq
Na_2S	s	-364.8	-349.8	83.7	2.53×10^{-1}	C dlq releases H$_2$S

† Uncertain. ‡ Highly uncertain. ^{dec} Decomposes.

5

Compound	State	Crystal system	M $\mathrm{g\,mol^{-1}}$	ρ $\mathrm{g\,cm^{-3}}$	T_m K	T_b K
Sodium (continued)						
Na_2SO_4	s	MCL	142.0	—	1157	—
$Na_2SO_4 \cdot 10H_2O$	s	MCL	322.2	1.46	306	373[dhd]
$NaHSO_4$	s	TCL	120.1	2.43	588	dec
$Na_2S_2O_3$	s	MCL	158.1	1.67	—	—
$Na_2S_2O_3 \cdot 5H_2O$	s	MCL	248.2	1.73	313–318	373[dhd]
Na_3PO_4	s	—	164.1	—	—	—
Na_2SiO_3 (water glass)	s	ORH	122.1	2.4	1361	—
$Na_2B_4O_7$	s	ORH	201.2	2.37	1014	1848[dec]
$Na_2B_4O_7 \cdot 10H_2O$	s	MCL	381.4	1.73	348	593[dhd]
$NaNH_2$ (amide)	s	ORH	39.0	—	483	673
Na^+	g	—	—	—	—	—
Strontium						
SrF_2	s	FCC	125.6	4.24	1746	2762
$SrCl_2$	s	CUB	158.5	3.05	1148	1523
$SrCl_2 \cdot H_2O$	s	—	176.5	—	—	—
$SrCl_2 \cdot 2H_2O$	s	MCL	194.6	2.67	—	—
$SrCl_2 \cdot 6H_2O$	s	HEX	266.6	1.93	388[dhd 4]	—
$Sr(ClO_4)_2$	s	CUB	286.5	—	—	—
$SrBr_2$	s	TET	247.4	4.22	916	dec
SrI_2	s	—	341.4	4.55	788	dec
$SrI_2 \cdot H_2O$	s	—	359.4	—	—	—
$SrI_2 \cdot 2H_2O$	s	—	377.4	—	—	—
$SrI_2 \cdot 6H_2O$	s	HEX	449.5	2.67	363.1[dec]	—
$Sr(IO_3)_2$	s	TRI	437.4	5.04	—	—
SrO	s	FCC	103.6	4.7	2703	3273[†]
$Sr(OH)_2$	s	ORH	121.6	3.63	648[in H2]	983[dhd 1]
$Sr(OH)_2 \cdot 8H_2O$	s	TET	265.8	1.90	373[dhd]	—
$SrCO_3$	s	ORH	147.6	3.70	1770[69 atm]	1613[−CO2]
$Sr(HCO_3)_2$	aq	—	209.6	—	—	—
$Sr(NO_3)_2$	s	CUB	211.6	2.99	843	1373
$Sr(NO_3)_2 \cdot 4H_2O$	s	MCL	283.7	2.2	373[dhd 4]	1373[dec]
SrS	s	CUB	119.7	3.70	>2273	—
$SrSO_4$	s	ORH	183.7	3.96	1878	—
Sr^{2+}	g	—	—	—	—	—
Sulphur						
SF_4	g	—	108.1	—	149	233
SF_6	g	—	146.1	1.88[223 K]	223[m]	209[sub]
SCl_2	g	—	103.0	1.62	195	332[dec]
SCl_4	l	—	173.9	—	243	258[dec]

[†] Uncertain.   [‡] Highly uncertain.   [sub] Sublimes.   [dec] Decomposes.   [dhd(n)] Dehydrates (loses n molecules of H_2O).   [m] Melts under pressure.

Compound	State	$\Delta H_f^\ominus$ kJ mol^{-1}	$\Delta G_f^\ominus$ kJ mol^{-1}	$S^\ominus$ J mol^{-1} K^{-1}	m_{sat} mol/100 g	Notes (see Table 5.1)
Sodium (continued)						
Na_2SO_4	s	−1387.1	−1270.2	149.6	3.03×10^{-2}	—
$Na_2SO_4 \cdot 10H_2O$	s	−4327.3	−3647.4	592.0	1.97×10^{-1}	eff Glaubers salt
$NaHSO_4$	s	−1125.5	−992.9	113.0	2.38×10^{-1}	C
$Na_2S_2O_3$	s	−1123.0	−1028.0	155.0	3.16×10^{-1}	—
$Na_2S_2O_3 \cdot 5H_2O$	s	−2607.9	−2230.1	372.4	4.80×10^{-1}	eff hypo
Na_3PO_4	s	−1917.4	−1788.9	173.8	—	—
Na_2SiO_3	s	−1554.9	−1461.0	113.8	8.39×10^{-1}	dlq
$Na_2B_4O_7$	s	−3291.1	−3096.2	189.5	5.27×10^{-3}	—
$Na_2B_4O_7 \cdot 10H_2O$	s	−6288.6	−5516.6	585.8	1.60×10^{-2}	borax
$NaNH_2$	s	−123.8	−64.0	76.9	dec W	C
Na^+	g	609.0	—	147.9	—	—
Strontium						
SrF_2	s	−1216.3	−1164.8	82.1	9.50×10^{-5}	P
$SrCl_2$	s	−828.9	−781.2	114.9	1.00×10^{-2}	
$SrCl_2 \cdot H_2O$	s	−1136.8	−1036.4	172.0	—	
$SrCl_2 \cdot 2H_2O$	s	−1438.0	−1282.0	218.0	—	transparent
$SrCl_2 \cdot 6H_2O$	s	−2623.8	−2241.2	390.8	—	
$Sr(ClO_4)_2$	s	−762.8	—	247.7	1.082	hyg
$SrBr_2$	s	−717.6	−697.1	135.1	4.33×10^{-1} (0.355)	
SrI_2	s	−558.1	−562.3	159.0	0.484	
$SrI_2 \cdot H_2O$	s	−886.6	—	—	—	
$SrI_2 \cdot 2H_2O$	s	−1182.4	—	—	—	
$SrI_2 \cdot 6H_2O$	s	−2388.6	—	—	—	yl dlq
$Sr(IO_3)_2$	s	−1019.2	−855.2	234.0	6.86×10^{-5}	
SrO	s	−592.0	−561.9	54.4	8.27×10^{-3}	
$Sr(OH)_2$	s	−959.0	−869.4	88.0	3.37×10^{-3}	dlq
$Sr(OH)_2 \cdot 8H_2O$	s	−3352.2	—	—	6.55×10^{-3}	dlq
$SrCO_3$	s	−1220.1	−1140.4	97.1	7.38×10^{-6}	
$Sr(HCO_3)_2$	aq	−1927.9	−1731.3	150.6	5.68×10^{-4}	
$Sr(NO_3)_2$	s	−978.2	−780.1	194.6	1.55×10^{-1} $^{353\,K}$ (0.186)	
$Sr(NO_3)_2 \cdot 4H_2O$	s	−2154.8	−1730.7	369.0	0.213	
SrS	s	−453.1	−448.5	68.2	insoluble (dec)	
$SrSO_4$	s	−1453.1	−1341.0	117.0	7.11×10^{-5}	
Sr^{2+}	g	1790.6	—	164.6	—	—
Sulphur						
SF_4	g	−774.9	−731.4	291.9	dec	—
SF_6	g	−1209.0	−1105.4	291.7	3.70×10^{-3}	—
SCl_2	g	−19.7	—	282.2	—	—
SCl_4	l	−56.1	—	—	dec	yl-br

† Uncertain. ‡ Highly uncertain. ^{dec} Decomposes.

Compound	State	Crystal system	$\dfrac{M}{\text{g mol}^{-1}}$	$\dfrac{\rho}{\text{g cm}^{-3}}$	$\dfrac{T_{\text{m}}}{\text{K}}$	$\dfrac{T_{\text{b}}}{\text{K}}$
Sulphur (continued)						
S_2Cl_2	l	—	135.0	1.68	193	409
$SOCl_2$	l	—	119.0	1.66	168	352
SO_2Cl_2	l	—	135.0	1.67	219	342
SO_2	g	—	64.1	1.43$^{200\,\text{K}}$	200	263
SO_3	l	—	80.1	1.97	290	318
S	g	—	32.06	—	392	718
S_2	g	—	64.1	—	392	718
S_8	g	—	256.5	—	392	718
S^{2-}	aq	—	—	—	—	—

For CS_2, see Carbon.

Compound	State	Crystal system	$\dfrac{M}{\text{g mol}^{-1}}$	$\dfrac{\rho}{\text{g cm}^{-3}}$	$\dfrac{T_{\text{m}}}{\text{K}}$	$\dfrac{T_{\text{b}}}{\text{K}}$
Tin						
SnH_4 (stannane)	g	—	122.7	—	123	221
$SnCl_2$	s	ORH	189.6	3.95	519	925
$SnCl_2 \cdot 2H_2O$	s	MCL	225.6	2.71	311	dec
$SnCl_4$	l	—	260.5	2.23	240	387
$SnBr_2$	s	ORH	278.5	5.12	488	893
$SnBr_4$	s	MCL	438.3	3.34	304	475
$SnBr_4 \cdot 8H_2O$	s	—	582.4	—	—	—
SnI_2	s	—	372.5	5.28	593	990
SnO	s	TET	134.7	6.45	1353^{dec}	—
SnO_2 (cassiterite)	s	TET	150.7	6.95	1903	2123^{sub}
SnS	s	ORH	150.8	5.22	1155	1503
$Sn(SO_4)_2$	s	—	310.8	—	—	—
Sn^{2+}	g	—	—	—	—	—
Sn^{4+}	g	—	—	—	—	—
Titanium						
TiH_2	s	TET	49.9	3.9	673^{dec}	—
$TiCl_2$	s	HEX	118.8	3.13	sub H_2	—
$TiCl_3$	s	HEX	154.3	2.64	713^{dec}	—
$TiCl_4$	l	—	189.7	1.73	248	410
$TiBr_2$	s	—	207.7	4.31	> 773^{dec}	—
$TiBr_3$	s	HEX	287.6	—	—	—
$TiBr_4$	s	CUB	367.5	2.6	312	503
TiI_2	s	HEX	301.7	4.99	873	1273
TiI_4	s	CUB	555.5	4.3	423	650
TiO_2 (anatase)	s	TET	79.9	3.84	2103^{A}	
Ti_2O_3	s	HEX	143.8	4.6	2403^{dec}	—
Ti^{2+}	g	—	—	—	—	—
Ti^{4+}	g	—	—	—	—	—

[†] Uncertain.   [‡] Highly uncertain.   [A] Rutile form of TiO_2.   [dec] Decomposes.   [sub] Sublimes.

Compound	State	$\Delta H_f^\ominus$ kJ mol^{-1}	$\Delta G_f^\ominus$ kJ mol^{-1}	$S^\ominus$ J mol^{-1} K^{-1}	m_{sat} mol/100 g	Notes (see Table 5.1)
Sulphur (continued)						
S_2Cl_2	l	-59.4	4.2	—	dec	**C Pv**(1) yl-br fuming
$SOCl_2$	l	-245.6	-197.9	307.9	dec	**C Pv** rd-br
SO_2Cl_2	l	-394.1	-305.0	216.7	dec	**C Pv** rd-br
SO_2	g	-296.8	-300.2	248.1	$1.66 \times 10^{-1\ 293\,K}$	**C Pg**(10) rd-br
SO_3	l	-441.0	-368.4	95.6	∞	dlq
S	g	278.8	238.3	167.8	—	—
S_2	g	128.4	79.3	228.1	—	—
S_8	g	102.3	49.7	430.9	—	—
S^{2-}	aq	33.1	85.8	-14.6	—	—

For CS$_2$, see Carbon.

Compound	State	$\Delta H_f^\ominus$ kJ mol^{-1}	$\Delta G_f^\ominus$ kJ mol^{-1}	$S^\ominus$ J mol^{-1} K^{-1}	m_{sat} mol/100 g	Notes (see Table 5.1)
Tin						**All Sn compounds P**
SnH_4	g	162.8	188.3	227.6	—	**P**
$SnCl_2$	s	-325.1	—	—	$1.42^{228\,K}$ (0.703)	**P**
$SnCl_2 \cdot 2H_2O$	s	-921.3	-787.8	—	dec	**P**
$SnCl_4$	l	-511.3	-440.2	258.6	soluble (dec)	**P**
$SnBr_2$	s	-243.5	-250.6	146.0	soluble (dec)	**P** yl
$SnBr_4$	s	-377.4	-350.2	264.4	dec	**P** dlq
$SnBr_4 \cdot 8H_2O$	s	-276.8	—	—	—	**P**
SnI_2	s	-143.5	-145.2	168.6	2.63×10^{-3}	**P**
SnO	s	-285.8	-256.9	56.5	5.00×10^{-7}	**P** bk
SnO_2	s	-580.7	-519.7	52.3	1.40×10^{-11}	**P**
SnS	s	-100.0	-98.3	77.0	1.3×10^{-8}	**P** gn-bk
$Sn(SO_4)_2$	s	-1629.2	-1443.0	155.2	—	**P**
Sn^{2+}	g	2434.9	—	168.4	—	—
Sn^{4+}	g	9323.2	—	168.4	—	—
Titanium						
TiH_2	s	-119.7	-80.3	29.1	—	gy
$TiCl_2$	s	-513.8	-464.4	87.4	dec	—
$TiCl_3$	s	-720.9	-653.5	139.7	soluble	—
$TiCl_4$	l	-804.2	-737.2	252.3	dec	—
$TiBr_2$	s	-402.0	-375.0	130.1	soluble$^{-H_2}$	bk
$TiBr_3$	s	-548.5	-523.8	176.6	—	—
$TiBr_4$	s	-616.7	-589.5	243.5	dec	—
TiI_2	s	-263.0	-270.1	147.7	dec	bk, hyg
TiI_4	s	-375.7	-371.5	249.4	v. soluble	rd
TiO_2	s	-939.7	-884.5	49.9	insoluble	—
Ti_2O_3	s	-1520.9	-1434.3	78.9	insoluble	—
Ti^{2+}	g	2450.6	—	—	—	—
Ti^{4+}	g	9290.2	—	—	—	—

† Uncertain. ‡ Highly uncertain. ^{dec} Decomposes.

Compound	State	Crystal system	$\dfrac{M}{\text{g mol}^{-1}}$	$\dfrac{\rho}{\text{g cm}^{-3}}$	$\dfrac{T_m}{K}$	$\dfrac{T_b}{K}$
Tungsten						
WF_6	l	—	297.8	3.44	276	291
WCl_2	s	AMS	254.8	5.44	—	—
WCl_4	s	—	325.7	4.62	dec	—
WCl_6	s	HEX	396.6	3.52	548	620
WBr_6	s	—	663.3	6.9	505	—
WO_3 (wolframite)	s	MCL	231.8	7.16	1746	—
WS_2	s	HCP	248.0	7.5	1523^{dec}	—
WC	s	HEX	195.9	15.63	3143	6273
W^+	g	—	—	—	—	—
Uranium						
UF_6	g	—	352.0	4.68	338	329
UF_2O_2	s	—	308.0	—	—	—
UCl_2	s	—	309.0	—	—	—
UCl_2O_2	s	—	340.9	—	851	dec
UO_2	s	CUB	270.0	10.96	3151	—
UO_3	s	ORH	286.0	7.29	dec	—
U_2C_3	s	—	512.1	—	—	—
$UO_2(NO_3)_2$	s	—	394.0	—	—	—
$UO_2(NO_3)_2 \cdot 6H_2O$	s	ORH	502.1	2.81	333	391
US_2	s	TET	302.2	7.96	>1373	ox
Vanadium						
VF_4	s	—	126.9	2.97	598^{dec}	—
VF_5	l	—	145.9	2.18	—	384
VF_5	g	—	145.9	—	—	—
VCl_2	s	—	121.8	3.23	—	—
VCl_3	s	HEX	157.3	3.00	dec	—
VCl_4	l	—	192.8	1.82	245	422
VBr_2	s	HEX	210.8	—	—	—
VBr_3	s	—	290.7	4.00	dec	—
VBr_4	g	—	370.6	—	—	—
VI_2	s	HEX	304.7	5.44	$1023–1073^{sub}$	—
VI_3	s	—	431.7	—	—	—
VI_4	g	—	558.6	—	—	—
VO	s	CUB	66.9	6.76	ign	—
V_2O_3	s	HEX	149.9	4.87	2243	—
VO_2	s	MCL	82.9	4.34	2240	—
V_2O_5	s	ORH	181.9	3.36	963	2023^{dec}
V^{2+}	g	—	—	—	—	—
V^{3+}	g	—	—	—	—	—
V^{4+}	g	—	—	—	—	—

[†] Uncertain.   [‡] Highly uncertain.   [ign] Ignites.   [dec] Decomposes.   [sub] Sublimes.

Compound	State	$\Delta H_f^{\ominus}$ kJ mol^{-1}	$\Delta G_f^{\ominus}$ kJ mol^{-1}	$S^{\ominus}$ J mol^{-1} K^{-1}	m_{sat} mol/100 g	Notes (see Table 5.1)
Tungsten						
WF$_6$	l	-1747.7	-1631.4	251.5	dec	yl
WCl$_2$	s	-255.0	-213.6	130.2	dec	gy
WCl$_4$	s	-467.0	-303.1	344.5	dec	—
WCl$_6$	s	-682.5	-548.9	254.0	dec	dk bl
WBr$_6$	s	-348.5	-328.0	472.0	insoluble	bk
WO$_3$	s	-842.9	-764.1	75.9	insoluble	yl
WS$_2$	s	-209.0	—	84.0	insoluble	dk gr, br
WC	s	-40.5	-40.2	35.6	insoluble	bk
W$^+$	g	1625.9	—	—	—	—
Uranium						
UF$_6$	g	-2112.9	-2029.2	379.7	dec	R P dlq
UF$_2$O$_2$	s	-1653.0	—	135.6	—	R P
UCl$_2$	s	-75.3	-80.3	79.0	—	R P
UCl$_2$O$_2$	s	-1263.1	-1159.0	150.5	0.939	R P yl, dlq
UO$_2$	s	-1129.7	-1075.3	77.8	3.00×10^{-7}	R P br-bk
UO$_3$	s	-1263.6	-1184.1	98.6	3.95×10^{-6}	R P yl-rd
U$_2$C$_3$	s	-205.0	-201.0	105.0	—	R P
UO$_2$(NO$_3$)$_2$	s	-1377.4	-1142.7	276.1	—	R P
UO$_2$(NO$_3$)$_2$·6H$_2$O	s	-3197.8	-2615.0	505.6	3.22×10^{-1}	R P dlq yl
US$_2$	s	-502.0	-531.7	110.5	—	R P gy-bk
Vanadium						
VF$_4$	s	-1403.3	—	—	soluble	yl
VF$_5$	l	-1480.3	-1373.2	175.7	—	—
VF$_5$	g	-1433.8	-1369.8	320.8	—	—
VCl$_2$	s	-452.0	-406.0	97.1	soluble, dec	gn dlq
VCl$_3$	s	-580.7	-511.3	131.0	soluble, dec	pk dlq
VCl$_4$	l	-569.4	-503.7	255.2	soluble, dec	rd-br
VBr$_2$	s	-365.3	—	126.0	—	—
VBr$_3$	s	-433.5	—	142.0	soluble	gn-bk, dlq
VBr$_4$	g	-336.8	—	335.0	—	—
VI$_2$	s	-251.5	—	143.1	soluble	vi-rose
VI$_3$	s	-270.7	—	215.5	—	—
VI$_4$	g	-122.6	—	—	—	—
VO	s	-431.8	-404.2	38.9	insoluble	gy
V$_2$O$_3$	s	-1228.0	-1139.3	98.3	insoluble	bk
VO$_2$	s	—	—	—	insoluble	bl
V$_2$O$_5$	s	-1550.6	-1419.6	131.0	soluble	or
V^{2+}	g	2590.5	—	169.4	—	—
V^{3+}	g	5430.5	—	171.5	—	—
V^{4+}	g	9943.3	—	169.3	—	—

† Uncertain. ‡ Highly uncertain. ^{dec} Decomposes.

Compound	State	Crystal system	$\dfrac{M}{\text{g mol}^{-1}}$	$\dfrac{\rho}{\text{g cm}^{-3}}$	$\dfrac{T_m}{K}$	$\dfrac{T_b}{K}$
Xenon						
XeF_2	s	TET	169.3	4.3	413[†]	—
XeF_4	s	MCL	207.2	4.1	387[†]	—
XeF_6	s	MCL	245.3	—	319	—
XeO_3	s	ORH	179.3	4.6	dec	—
Xe^{2+}	g	—	—	—	—	—
Xe^{3+}	g	—	—	—	—	—
Zinc						
ZnF_2	s	TET	103.4	4.95	1145	1773
$ZnCl_2$	s	HEX	136.3	2.91	556	1005
$ZnBr_2$	s	HEX	225.2	4.20	667	923
ZnI_2	s	HEX	319.2	4.74	719	897
ZnO (zincite)	s	HEX	81.4	5.61	2248	—
$ZnCO_3$	s	HEX	125.4	4.40	573[dec]	—
$Zn(NO_3)_2$	s	—	189.4	—	—	—
$Zn(NO_3)_2 \cdot 6H_2O$	s	ORH	297.5	2.06	309	378–404[dhd 6]
ZnS (wurtzite)	s	HEX	97.4	3.98	1973[50 atm]	1458[sub]
ZnS (blende)	s	HEX	97.4	4.10	1293[tr]	—
$ZnSO_4$	s	ORH	161.4	3.54	873[dec]	—
$ZnSO_4 \cdot 7H_2O$	s	ORH	287.5	1.96	373	553[dhd]
Zn^{2+}	g	—	—	—	—	—

[†] Uncertain. [‡] Highly uncertain. [dec] Decomposes. [sub] Sublimes. [tr] Transition. [dhd] Dehydrates.

5·4 | ORGANIC COMPOUNDS: SOME TRADITIONAL AND SYSTEMATIC NAMES

Traditional name	Recommended name	Traditional name	Recommended name
acetaldehyde	ethanal	aspartic acid	aminobutanedioic acid
acetamide	ethanamide	azobenzene	(phenylazo)benzene
acetanilide	N-phenylethanamide	benzal chloride	(dichloromethyl)benzene
acetate	ethanoate	benzyl chloride	(chloromethyl)benzene
acetic acid	ethanoic acid	butylamine	1-aminobutane
acetone	propanone	butyraldehyde	butanal
acetonitrile	ethanenitrile	butyric acid	butanoic acid
acetophenone	phenylethanone	carbon tetrachloride	tetrachloromethane
acetyl	ethanoyl	catechol	benzene-1,2-diol
acetylene	ethyne	chloral	trichloroethanol
acrolein	propenal	chloroform	trichloromethane
acrylic acid	propenoic acid	ethyl methyl ether	methoxymethane
adipic acid	hexanedioic acid	ether (diethyl ether)	ethoxymethane
alanine	2-aminopropanoic acid	ethyl acetoacetate	ethyl 3-oxobutanoate
alcohol (ethyl)	ethanol	ethyl iodide	iodoethane
alcohol (wood)	methanol	ethyl methyl ketone	butanone
aniline	phenylamine	ethylene	ethene
anisole	methoxybenzene	ethylenediamine	ethane-1,2-diamine

Compound	State	$\Delta H_f^{\ominus}$ kJ mol^{-1}	$\Delta G_f^{\ominus}$ kJ mol^{-1}	$S^{\ominus}$ J mol^{-1} K^{-1}	m_{sat} mol/100 g	Notes (see Table 5.1)
Xenon						
XeF$_2$	s	-133.9	-62.8	133.9	hyd	—
XeF$_4$	s	-261.5	-121.3	146.4	hyd	—
XeF$_6$	s	-380.7	—	—	hyd	—
XeO$_3$	s	401.7	—	—	—	—
Xe^{2+}	g	3229.2	—	—	—	—
Xe^{3+}	g	6335.0	—	—	—	—
Zinc						
ZnF$_2$	s	-764.4	-449.5	73.7	1.57×10^{-2}	—
ZnCl$_2$	s	-415.1	-369.4	111.5	3.03	hyg
ZnBr$_2$	s	-328.7	-312.1	138.5	2.09	dlq
ZnI$_2$	s	-208.0	-208.9	161.1	1.35	—
ZnO	s	-348.3	-318.3	43.6	1.23×10^{-5}	—
ZnCO$_3$	s	-812.8	-731.6	82.4	1.64×10^{-4}	—
Zn(NO$_3$)$_2$	s	-483.7	—	—	—	—
Zn(NO$_3$)$_2$·6H$_2$O	s	-2306.6	-1773.1	456.9	0.620	—
ZnS (wurtzite)	s	-192.6	-187.0	57.7	1.47×10^{-10}	—
ZnS (blende)	s	-206.0	-201.3	65.3	—	—
ZnSO$_4$	s	-982.8	-874.5	119.7	soluble	—
ZnSO$_4$·7H$_2$O	s	-3077.8	-2563.1	388.7	3.56×10^{-1}	eff
Zn^{2+}	g	2782.7	—	160.9	—	—

References: American Society for testing materials, Gray, Linke, Pieters, Stephen, Stull, Wagman, Weast.

5

ORGANIC COMPOUNDS: SOME TRADITIONAL AND SYSTEMATIC NAMES | 5·4 |

Traditional name	Recommended name	Traditional name	Recommended name
ethylene dibromide	1,2-dibromoethane	*o*-cresol	2-methylphenol
ethylene glycol	ethane-1,2-diol	olefins	alkenes
ethylene oxide	epoxyethane	oxalic acid	ethanedioic acid
fatty acids	alkanoic acids	*o*-xylene	1,2-dimethylbenzene
formaldehyde	methanal	phosgene	carbonyl chloride
formate	methanoate	phthalic acid	benzene-1,2-dicarboxylic acid
formic acid	methanoic acid	propionaldehyde	propanal
glycerine } glycerol }	propane-1,2,3-triol	sec-butyl	1-methylpropyl
		stearic acid	octadecanoic acid
glycine	aminoethanoic acid	styrene	phenylethene
glycols	diols	succinic acid	butanedioic acid
isobutyl	2-methylpropyl	tert-butyl	1,1-dimethylethyl
isobutyric acid	methylpropanoic acid	thiourea	thiocarbamide
isopropyl	1-methylethyl	toluene	methylbenzene
lactic acid	2-hydroxypropanoic acid	urea	carbamide
lauryl alcohol	dodecan-1-ol	vinyl acetate	ethenyl ethanoate
lauroyl peroxide	di(dodecanoyl) peroxide	vinyl chloride	chloroethene
m-xylene	1,3-dimethylbenzene	*Reference:* ASE (1984).	

See Table 5.1 for general notes and abbreviations, and Table 5.4 for some old organic names and their systematic equivalents.

St State: s solid; l liquid; g gas;
   aq aqueous.
M Molar mass.

ρ Density at 298 K or density of liquid at just below T_b for gases, unless otherwise indicated.
n Refractive index at 298 K or just below T_b for gases.
T_m Melting temperature ⎰at 1 atm except where
T_b Boiling temperature ⎱otherwise stated.

Compound	Formula	St	M $\mathrm{g\,mol^{-1}}$	ρ $\mathrm{g\,cm^{-3}}$	n	T_m K	T_b K
Carbon monoxide	CO	g	28.0	$1.25 \times 10^{-3\,gas}$		74.1	81.6
Carbon dioxide	CO_2	g	44.0	$1.98 \times 10^{-3\,gas}$		216.5^A	194.0^{sub}
STRAIGHT CHAIN ALKANES							
Methane	CH_4	g	16.0	0.466^{liq}	—	91.1	109.1
Ethane	CH_3CH_3	g	30.1	0.572^{liq}	—	89.8	184.5
Propane	$CH_3CH_2CH_3$	g	44.1	0.585^{liq}	—	83.4	231.0
Butane	$CH_3(CH_2)_2CH_3$	g	58.1	0.601^{liq}	1.3326	134.7	272.6
Pentane	$CH_3(CH_2)_3CH_3$	l	72.2	0.626	1.3575	143.1	309.2
Hexane	$CH_3(CH_2)_4CH_3$	l	86.2	0.660	1.3749	178.1	342.1
Heptane	$CH_3(CH_2)_5CH_3$	l	100.2	0.684	1.3876	182.5	371.5
Octane	$CH_3(CH_2)_6CH_3$	l	114.2	0.703	1.3974	216.3	398.8
Nonane	$CH_3(CH_2)_7CH_3$	l	128.3	0.718	1.4054	222.1	423.9
Decane	$CH_3(CH_2)_8CH_3$	l	142.3	0.730	1.4119	243.4	447.2
Undecane	$CH_3(CH_2)_9CH_3$	l	156.3	0.740	1.4398	247.5	469.1
Dodecane	$CH_3(CH_2)_{10}CH_3$	l	170.3	0.749	1.4216	263.5	489.4
Eicosane	$CH_3(CH_2)_{18}CH_3$	s	282.6	0.789	1.4405	309.9	616.9
BRANCHED ALKANES							
2-Methylpropane	$(CH_3)_2CHCH_3$	g	58.1	0.557^{liq}	—	113.7	261.4
2-Methylbutane	$(CH_3)_2CHCH_2CH_3$	l	72.2	0.620	1.3537	113.2	301.0
2-Methylpentane	$(CH_3)_2CH(CH_2)_2CH_3$	l	86.2	0.653	1.3715	119.4	333.4
2-Methylhexane	$(CH_3)_2CH(CH_2)_3CH_3$	l	100.2	0.679	1.3848	154.8	363.1
2-Methylheptane	$(CH_3)_2CH(CH_2)_4CH_3$	l	114.2	0.698	1.3949	164.1	390.7
2,2-Dimethylpropane	$C(CH_3)_4$	g	72.2	0.591^{liq}	1.3420	256.6	282.6
CYCLO-ALKANES							
Cyclopropane	$(CH_2)_3$	g	42.1	—	—	145.5	240.4
Cyclobutane	$(CH_2)_4$	g	56.1	0.694^{liq}	1.3650	182.4	285.1

^{A} At 5.2 atm. ^{liq} Liquid. ^{sub} Sublimes.

Enthalpy changes of formation of gaseous ions containing carbon

Ion(g)	CO^+	CO^{2+}	CO_2^+	CH^+	CH_2^+	CH_3^+	CH_4^+
$\Delta H_f^{\ominus}/\mathrm{kJ\,mol^{-1}}$	1247.5	3956.0	942.4	1675.3	1401.2	1095.0	1157.7

See Table 5.1 for general notes and abbreviations; and see lefthand page for state.

$\Delta H_c^{\ominus}$ Standard molar enthalpy change of combustion at 298 K.[A]

$\Delta H_f^{\ominus}$ Standard molar enthalpy change of formation at 298 K.

$\Delta G_f^{\ominus}$ Standard molar Gibbs free energy change of formation at 298 K.

$S^{\ominus}$ Standard molar entropy at 298 K.

Chosen standard pressure is 1 atm.

p Dipole moment in the gas phase; chosen unit is the common non-SI unit the debye ($D \triangleq 3.34 \times 10^{-30}$ C m).

ε_r Relative permittivity (static, 298 K), 'dielectric constant'.

Notes See Table 5.1 for abbreviations used.

Compound	$\Delta H_c^{\ominus}$ kJ mol^{-1}	$\Delta H_f^{\ominus}$ kJ mol^{-1}	$\Delta G_f^{\ominus}$ kJ mol^{-1}	$S^{\ominus}$ J mol^{-1}K^{-1}	$\dfrac{p}{D}$	ε_r	Notes
Carbon monoxide	-283.0	-110.5	-137.2	197.6	—	—	Pg(100)
Carbon dioxide	—	-393.5	-394.4	213.6	—	—	—
STRAIGHT CHAIN ALKANES							
Methane	-890.3	-74.8	-50.8	186.2	0	—	F
Ethane	-1559.7	-84.7	-32.9	229.5	0	—	F
Propane	-2219.2	-104.5	-23.4	269.9	0	1.66^{liq}	F
Butane	-2876.5	-126.5	-15.6	310.1	0	1.78^{liq}	F
Pentane	-3509.1	-173.2	-9.2	261.2	0	1.84	F
Hexane	-4163.0	-198.6	-4.2	295.9	0	1.89	F
Heptane	-4816.9	-224.0	$+1.3$	328.5	0	1.92	F
Octane	-5470.2	-250.0	$+6.4$	361.1	0	1.95	F
Nonane	-6124.6	-274.9	$+11.9$	393.7	0	1.97	F
Decane	-6777.9	-300.9	$+17.4$	425.9	0	—	F
Undecane	-7430.9	-327.2	$+22.8$	—	0	—	F
Dodecane	-8086.5	-350.9	$+28.4$	—	0	—	F
Eicosane	—	—	—	—	0	2.08	F
BRANCHED ALKANES							
2-Methylpropane	-2868.5	-134.5	-17.9	294.6	0	1.73	—
2-Methylbutane	-3503.4	-178.9	-14.5	260.4	—	1.84	Isopentane
2-Methylpentane	-4157.0	-204.6	-8.1	—	—	—	—
2-Methylhexane	-4811.4	-229.5	-2.0	—	—	—	—
2-Methylheptane	-5465.2	-255.0	$+3.8$	—	—	—	—
2,2-Dimethylpropane	-3492.5	-189.8	-15.2	306.4	—	1.80	Neopentane
CYCLO-ALKANES							
Cyclopropane	-2091.4	$+53.3$	$+104.1$	—	0	—	E
Cyclobutane	-2720.9	$+3.7$	—	—	—	—	—

^{liq} Values for ε_r marked liq relate to undercooled liquid.

[A] *Enthalpy changes of combustion*, $\Delta H_c^{\ominus}$. These refer to the production of CO_2(g), N_2(g), H_2O(l), HCl (600H$_2$O), Br$_2$(l), I$_2$(s), and H$_2$SO$_4$ (115H$_2$O). It is impracticable to include fluorine compounds in the list because the dilution states of HF and the amount of CF$_4$ produced vary greatly from case to case.

For a compound of general formula $C_aH_bO_cN_d$ in a particular state, the enthalpy change of combustion corresponds to the process:

$$C_aH_bO_cN_d + (a + \tfrac{1}{4}b - \tfrac{1}{2}c)O_2(g) \rightarrow$$
$$aCO_2(g) + \tfrac{1}{2}bH_2O(l) + \tfrac{1}{2}dN_2(g) \text{ at } 298 \text{ K}$$

hence: $\Delta H_c^{\ominus} = a\,\Delta H_f^{\ominus}(CO_2)(g) + \tfrac{1}{2}b\,\Delta H_f^{\ominus}(H_2O)(l)$
$$- \Delta H_f^{\ominus}(C_aH_bO_cN_d)$$
$$= -\Delta H_f^{\ominus}(C_aH_bO_cN_d) - (393.5a + 142.9b)\,\text{kJ mol}^{-1}$$
at 298 K and 1 atm.

Compound	Formula	St	$\dfrac{M}{\text{g mol}^{-1}}$	$\dfrac{\rho}{\text{g cm}^{-3}}$	n	$\dfrac{T_m}{K}$	$\dfrac{T_b}{K}$
CYCLO-ALKANES (continued)							
Cyclopentane	$CH_2(CH_2)_3CH_2$	l	70.1	0.745	1.4070	179.2	322.3
Cyclohexane	$CH_2(CH_2)_4CH_2$	l	84.2	0.779	1.4260	279.6	353.8
Cycloheptane	$CH_2(CH_2)_5CH_2$	l	98.2	0.810	1.4436	261.1	391.6
Cyclooctane	$CH_2(CH_2)_6CH_2$	l	112.2	0.835	1.4586	287.4	421.1
Cyclononane	$CH_2(CH_2)_7CH_2$	l	126.2	0.853	1.4328	282.7	444.0
Cyclodecane	$CH_2(CH_2)_8CH_2$	l	140.3	0.857	1.4714	282.5	474.0
ALKENES							
Ethene	$CH_2{=}CH_2$	g	28.1	0.610^{liq}	—	104.1	169.4
Propene	$CH_2{=}CHCH_3$	g	42.1	0.514^{liq}	—	87.9	225.7
But-1-ene	$CH_2{=}CHCH_2CH_3$	g	56.1	0.595^{liq}	—	87.8	266.8
trans-But-2-ene	$CH_3CH{=}CHCH_3$	g	56.1	0.604^{liq}	—	167.6	274.0
cis-But-2-ene	$CH_3CH{=}CHCH_3$	g	56.1	0.621^{liq}	—	134.2	276.8
Hex-1-ene	$CH_2{=}CH(CH_2)_3CH_3$	l	84.2	0.673	1.3880	133.3	336.4
Buta-1,2-diene	$CH_2{=}C{=}CHCH_3$	g	54.1	0.652^{liq}	—	136.9	283.9
Buta-1,3-diene	$CH_2{=}CHCH{=}CH_2$	g	54.1	0.621^{liq}	1.4290	164.5	268.7
Cyclohexene	$CH_2(CH_2)_3CH{=}CH$	l	81.2	0.811	1.4467	169.6	356.5
Phenylethene (styrene)	$C_6H_5CH{=}CH_2$	l	104.2	0.906	1.5470	242.5	418.3
ALKYNES							
Ethyne (acetylene)	$CH{\equiv}CH$	g	26.0	0.618^{liq}	1.0005	192.3	189.1
Propyne	$CH_3C{\equiv}CH$	g	40.1	0.671^{liq}	—	171.6	249.9
1-Butyne	$CH_3CH_2{\equiv}CH$	g	54.1	0.678^{liq}	1.3962	147.4	281.2
2-Butyne	$CH_3C{\equiv}CCH_3$	l	54.1	0.691	1.3921	240.9	300.1
ARENES							
Benzene	C_6H_6	l	78.1	0.879	1.5010	278.6	353.2
Naphthalene	$C_{10}H_8$	s	128.2	1.101	1.5898^B	353.6	491.1
Methylbenzene (toluene)	$C_6H_5CH_3$	l	92.1	0.867	1.4970	178.1	383.7
Ethylbenzene	$C_6H_5CH_2CH_3$	l	106.2	0.867	1.4960	178.1	409.3
Propylbenzene	$C_6H_5(CH_2)_2CH_3$	l	120.2	0.862	1.4920	173.6	432.3
1,2-Dimethylbenzene	$C_6H_4(CH_3)_2$	l	106.2	0.880	1.5060	247.9	417.5
1,3-Dimethylbenzene	$C_6H_4(CH_3)_2$	l	106.2	0.864	1.4970	225.2	412.2
1,4-Dimethylbenzene	$C_6H_4(CH_3)_2$	l	106.2	0.861	1.4960	286.4	411.4
Ethenylbenzene (styrene)	$C_6H_5CH{=}CH_2$	l	104.2	0.906	1.5470	242.5	418.3
AMINES (AMINOALKANES, ETC.)							
Methylamine	CH_3NH_2	g	31.1	0.660^{liq}	1.3527	179.6	266.8
Dimethylamine	$(CH_3)_2NH$	g	45.1	0.656^{liq}	1.3597	180.1	280.5
Trimethylamine	$(CH_3)_3N$	g	59.1	0.633^{liq}	1.3476	155.9	276.0
Ethylamine	$CH_3CH_2NH_2$	g	45.1	0.683^{liq}	1.3663	192.1	289.7
1-Aminopropane	$CH_3CH_2CH_2NH_2$	l	59.1	0.717	1.3882	190.1	320.9
2-Aminopropane	$CH_3CHNH_2CH_3$	l	59.1	0.688	1.3742	177.9	305.5
1-Aminobutane	$CH_3(CH_2)_3NH_2$	l	73.1	0.739	1.4014	224.0	350.9
2-Aminobutane	$CH_3CH_2CHNH_2CH_3$	l	73.1	0.734	1.3972	$<201.1^†$	336.6

^{liq} Liquid.　^{B} At 358 K.　† Uncertain.

Compound	$\dfrac{\Delta H_c^\ominus}{\text{kJ mol}^{-1}}$	$\dfrac{\Delta H_f^\ominus}{\text{kJ mol}^{-1}}$	$\dfrac{\Delta G_f^\ominus}{\text{kJ mol}^{-1}}$	$\dfrac{S^\ominus}{\text{J mol}^{-1}\text{K}^{-1}}$	$\dfrac{p}{D}$	ε_r	Notes
CYCLO-ALKANES (continued)							
Cyclopentane	−3289.4	−107.1	+36.5	204.3	0	1.97[A]	—
Cyclohexane	−3919.5	−156.3	+26.8	204.4	—	2.02	—
Cycloheptane	−4598.4	−156.7	—	—	—	—	—
Cyclooctane	−5266.7	−167.7	—	—	—	—	—
Cyclononane	−5932.5	−181.2	—	—	—	—	—
Cyclodecane	−6586.3	−206.7	—	—	—	—	—
ALKENES							
Ethene	−1410.8	+52.2	+68.2	219.5	0	—	Ethylene
Propene	−2058.1	+20.2	+74.7	266.9	0.35	1.86	—
But-1-ene	−2716.8	−0.4	+72.0	305.6	0.38	—	—
trans-But-2-ene	−2705.0	−12.2	+62.9	296.4	0	—	—
cis-But-2-ene	−2709.4	−7.8	+65.9	300.8	—	—	—
Hex-1-ene	−4003.4	−72.4	—	—	—	—	—
Buta-1,2-diene	−2593.7	+162.3	+201.5	293.0	—	—	—
Buta-1,3-diene	−2541.3	+109.9	+151.9	278.7	0	—	—
Cyclohexene	−3751.9	−38.1	—	—	0.55	2.22	—
Phenylethene	−4395.0	+103.8	+202.5	345.1	0	2.43	—
ALKYNES							
Ethyne	−1300.8	+228.0	+209.2	200.8	0	—	E
Propyne	−1938.7	+186.6	+194.2	248.1	0.75	—	E
1-Butyne	−2596.6	+165.2	+203.1	—	—	—	—
2-Butyne	−2576.8	+118.8	+187.2	—	—	1.39	—
ARENES							
Benzene	−3267.4	+49.0	+124.5[†]	172.8[†]	0	2.28	**Ps Pv**(25) **F**
Naphthalene	−5155.9	+77.7	—	—	—	2.54	—
Methylbenzene	−3909.8	+12.1	+110.6	319.7	0.36	2.38	—
Ethylbenzene	−4563.9	−13.1	+119.7	255.2[†]	0.35	2.24	—
Propylbenzene	−5218.0	−38.3	+123.8	290.5[†]	—	2.27	—
1,2-Dimethylbenzene	−4552.6	−24.4	+110.6	246.5	0.62	2.27	*o*-xylene
1,3-Dimethylbenzene	−4551.6	−25.4	+107.8	252.1	—	2.24	*m*-xylene
1,4-Dimethylbenzene	−4552.6	−24.4	+110.3	247.4	0	2.24	*p*-xylene
Ethenylbenzene	−4395.0	+103.8	+202.5	345.1	0	2.43	—
AMINES (AMINOALKANES, ETC.)							
Methylamine	−1085.0	−23.0	+32.1	243.3	1.30	9.4	**F**
Dimethylamine	−1768.8	−18.5	+59.2	280.5	0.93	5.26	**F**
Trimethylamine	−2442.9	−23.7	+76.7	287.0	0.71	5.5[B]	**F**
Ethylamine	−1739.8	−47.5	—	—	0.99	5.26	**F**
1-Aminopropane	−2365.1	−101.5	—	—	1.35	2.44	**C Pv**(5) **F**
2-Aminopropane	−2354.3	−112.3	—	—	—	—	**C Pv**(5) **F**
1-Aminobutane	−3018.3	−127.6	−81.8	—	1.32	—	—
2-Aminobutane	−3008.4	−137.5	—	—	—	—	—

[A] At 293 K. [B] At 360 MHz. [†] Uncertain.

5

Compound	Formula	St	$\dfrac{M}{g\,mol^{-1}}$	$\dfrac{\rho}{g\,cm^{-3}}$	n	$\dfrac{T_m}{K}$	$\dfrac{T_b}{K}$
AMINES (AMINOALKANES, ETC) (continued)							
Diethylamine	$(C_2H_5)_2NH$	l	73.1	0.706	1.3864	225.1	329.4
Triethylamine	$(C_2H_5)_3N$	l	101.2	0.728	1.4010	158.4	362.4
Phenylamine (aniline)	$C_6H_5NH_2$	l	93.1	1.022	1.5863	266.8	457.1
ORGANIC HALOGEN COMPOUNDS							
Fluoromethane	CH_3F	g	34.0	0.557^{liq}	1.1727	131.3	194.7
Chloromethane	CH_3Cl	g	50.5	0.916^{liq}	1.3390	176.0	248.9
Bromomethane	CH_3Br	g	94.9	1.676^{liq}	1.4218	179.5	276.7
Iodomethane	CH_3I	l	141.9	2.279	1.5308	206.7	315.5
Dichloromethane	CH_2Cl_2	l	84.9	1.316	1.4211	178.0	313.1
Trichloromethane	$CHCl_3$	l	119.4	1.479	1.4429	209.6	334.8
Tetrachloromethane	CCl_4	l	153.8	1.594	1.4601	250.1	349.6
Tetrabromomethane	CBr_4	s	331.6	3.420	—	$363-7^\dagger$	$462-3^\dagger$
Tetraiodomethane	CI_4	s	519.6	4.320	—	—	$403-13^{sub}$
Chloroethane	CH_3CH_2Cl	g	64.5	0.898^{liq}	1.3676	136.7	285.4
Bromoethane	CH_3CH_2Br	l	109.0	1.461	1.4239	154.5	311.5
Iodoethane	CH_3CH_2I	l	156.0	1.936	1.5133	165.1	345.4
1,2-Dibromoethane	CH_2BrCH_2Br	l	187.9	2.179	1.5387	282.9	404.4
1,2-Dichloroethane	$C_2H_4Cl_2$	l	99.0	1.235	1.4448	237.8	356.6
1,1,1-Trichloroethane	CH_3CCl_3	l	133.4	1.339	1.4379	242.7	347.2
Tetrachloroethene	C_2Cl_4	l	165.8	1.623	1.5030	254.1	394.1
1-Chloropropane	$CH_3CH_2CH_2Cl$	l	78.5	0.891	1.3879	150.3	319.7
2-Chloropropane	$CH_3CHClCH_3$	l	78.5	0.863	1.3777	155.9	308.8
1-Bromopropane	$CH_3CH_2CH_2Br$	l	123.0	1.354	1.4343	163.1	344.1
2-Bromopropane	$CH_3CHBrCH_3$	l	123.0	1.314	1.4250	184.1	332.5
1-Iodopropane	$CH_3CH_2CH_2I$	l	170.0	1.748	1.5058	172.1	375.5
2-Iodopropane	CH_3CHICH_3	l	170.0	1.703	1.5028	183.0	362.5
1-Chlorobutane	$CH_3(CH_2)_3Cl$	l	92.6	0.886	1.4021	150.0	351.5
1-Bromobutane	$CH_3(CH_2)_3Br$	l	137.0	1.276	1.4401	160.7	374.7
(+)-2-Bromobutane	$CH_3CH_2CHBrCH_3$	l	137.0	1.259	1.4367	161.2	364.3
1-Iodobutane	$CH_3(CH_2)_3I$	l	184.0	1.615	1.5001	170.1	403.6
2-Chloro-2-methylpropane	$(CH_3)_2CClCH_3$	l	92.6	0.842	1.3857	247.7	323.8
2-Bromo-2-methylpropane	$(CH_3)_2CBrCH_3$	l	137.0	1.221	1.4278	256.9	346.4
2-Iodo-2-methylpropane	$(CH_3)_2CICH_3$	l	184.0	1.571	—	234.9	373.1
Chlorobenzene	C_6H_5Cl	l	112.6	1.106	1.5241	227.5	405.1
Bromobenzene	C_6H_5Br	l	157.0	1.495	1.5597	242.3	429.1
Iodobenzene	C_6H_5I	l	204.1	1.831	1.5439	241.7	461.4
(Chloromethyl)benzene	$C_6H_5CH_2Cl$	l	126.6	1.102	—	234.1	452.4
ALCOHOLS							
Methanol	CH_3OH	l	32.0	0.793	1.3280	179.2	338.1
Ethanol	CH_3CH_2OH	l	46.1	0.789	1.3610	155.8	351.6
Propan-1-ol	$CH_3CH_2CH_2OH$	l	60.1	0.804	1.3860	146.6	370.5
Propan-2-ol	$CH_3CHOHCH_3$	l	60.1	0.787	1.3772	183.6	355.5

^{liq} Liquid. ^{sub} Sublimes. † Uncertain.

Compound	$\dfrac{\Delta H_c^{\ominus}}{\text{kJ mol}^{-1}}$	$\dfrac{\Delta H_f^{\ominus}}{\text{kJ mol}^{-1}}$	$\dfrac{\Delta G_f^{\ominus}}{\text{kJ mol}^{-1}}$	$\dfrac{S^{\ominus}}{\text{J mol}^{-1}\text{K}^{-1}}$	$\dfrac{p}{\text{D}}$	ε_r	Notes
AMINES (AMINOALKANES, ETC.) (continued)							
Diethylamine	−3042.1	−103.8	—	—	—	—	—
Triethylamine	−4376.8	−127.7	—	—	0.82	2.42	—
Phenylamine	−3392.6	+31.3	—	—	1.53	6.89	**Ps Pv**(5)
ORGANIC HALOGEN COMPOUNDS							
Fluoromethane	—	−247.0	−223.0	—	—	—	—
Chloromethane	−764.0	−82.0	−57.4	234.5	1.86	12.6[C]	—
Bromomethane	−769.9	−37.2	−25.9	246.3	1.79	9.82[D]	**Ps Pg**(20)
Iodomethane	−814.6	−15.5	+13.4	163.2	1.64	7.00	**Ps Pv**
Dichloromethane	−605.8	−124.1	−63.2	177.8	1.54	9.08	—
Trichloromethane	−474.0	−135.1	−71.4	201.8	1.02	4.81	**Pv**
Tetrachloromethane	−359.9	−129.6	−65.3	216.4	0	2.24	—
Tetrabromomethane	—	+18.8	+47.7	212.5	0	—	—
Tetraiodomethane	—	—	—	—	0	—	—
Chloroethane	−1413.1	−136.8	−52.9	—	1.98	—	**Ps Pv**(25)
Bromoethane	−1424.7	−90.5	—	—	2.02	—	**Ps Pv**(200)
Iodoethane	−1466.5	−40.7	—	—	1.90	—	—
1,2-Dibromoethane	—	−37.8[gas]	−80.7[gas]	—	1.40	—	**Ps Pv**(25)
1,2-Dichloroethane	−1246.4	−165.2	—	—	—	—	—
1,1,1-Trichloroethane	−1108.0	−177.3	—	—	—	—	—
Tetrachloroethene	−830.9	−48.6	—	—	—	—	—
1-Chloropropane	−2072.1	−161.3	—	—	2.10	—	—
2-Chloropropane	−2028.4	−172.2	—	—	2.04	—	—
1-Bromopropane	−2056.8	−116.4	—	—	1.93[†]	—	—
2-Bromopropane	−2052.0	−128.5	—	—	2.04[†]	—	—
1-Iodopropane	—	−68.4	—	—	1.74	—	—
2-Iodopropane	—	−75.7	—	—	1.95	—	—
1-Chlorobutane	−2704.1	−187.9	—	—	2.16	—	—
1-Bromobutane	−2716.5	−143.8	—	—	1.93	—	—
(+)-2-Bromobutane	−2705.2	−155.2	—	—	2.12	—	—
1-Iodobutane	—	—	—	—	1.88	—	—
2-Chloro-2-methylpropane	−2692.8	−191.1	—	—	2.13	—	—
2-Bromo-2-methylpropane	—	−163.4	—	—	—	—	—
2-Iodo-2-methylpropane	—	−107.4	—	—	—	—	—
Chlorobenzene	−3111.6	+11.0	+93.6	—	1.67	5.62	**Pv**(75) **F**
Bromobenzene	—	+60.5	+112.2	—	1.77	5.40	—
Iodobenzene	−3192.8	+114.5	+208.0	—	1.70	—	—
(Chloromethyl)benzene	−3708.7	−32.8	—	—	1.85	—	**C Pv**(1) **F**
ALCOHOLS							
Methanol	−726.0	−239.1	−166.4	239.7	1.70	32.6	**P**
Ethanol	−1367.3	−277.1	−174.9	160.7	1.69	24.3	—
Propan-1-ol	−2021.0	−302.7	−171.3	196.6	1.66	20.1	—
Propan-2-ol	−2005.8	−317.9	−180.3	180.5	1.68	18.1	—

[C] At 253 K.   [D] At 273 K.   [†] Uncertain.

Compound	Formula	St	$\dfrac{M}{\text{g mol}^{-1}}$	$\dfrac{\rho}{\text{g cm}^{-3}}$	n	$\dfrac{T_m}{\text{K}}$	$\dfrac{T_b}{\text{K}}$
ALCOHOLS (continued)							
Butan-1-ol	$CH_3(CH_2)_2CH_2OH$	l	74.1	0.810	1.3990	183.6	390.3
Pentan-1-ol	$CH_3(CH_2)_3CH_2OH$	l	88.2	0.815	1.4100	194.1	411.1
Hexan-1-ol	$CH_3(CH_2)_4CH_2OH$	l	102.2	0.820	1.4180	226.4	431.1
Heptan-1-ol	$CH_3(CH_2)_5CH_2OH$	l	116.2	0.822	1.4240	239.0	449.1
Octan-1-ol	$CH_3(CH_2)_6CH_2OH$	l	130.2	0.826	1.4295	256.4	467.5
Ethane-1,2-diol	CH_2OHCH_2OH	l	62.1	1.114	1.4318	261.6	471.1
Propane-1,2,3-triol	$CH_2OHCH_2OCH_2OH$	l	92.1	1.260	1.4746	293.1	563.1
2-Methylpropan-2-ol	$(CH_3)_3COH$	l	74.1	0.789	1.3878	298.6	355.4
Cyclohexanol	$CH_2(CH_2)_4CHOH$	s	100.2	0.962	1.463	298.2	434.2
ETHERS							
Methoxymethane	CH_3OCH_3	g	46.1	0.669	1.3018	134.6	248.1
Ethoxyethane	$CH_3CH_2OCH_2CH_3$	l	74.1	0.713	1.3524	156.9	307.6
Methoxybenzene (anisole)	$C_6H_5OCH_3$	l	108.1	0.994	—	235.6	428.1
ALDEHYDES							
Methanal (formaldehyde)	HCHO	g	30.0	0.815	—	181.1	252.1
Ethanal (acetaldehyde)	CH_3CHO	g	44.1	0.778	1.3311	152.1	293.9
Propanal	CH_3CH_2CHO	l	58.1	0.797	1.3619	192.1	321.9
Butanal	$CH_3CH_2CH_2CHO$	l	72.1	0.801	1.3791	174.1	348.8
2-Methylpropanal	$(CH_3)_2CHCHO$	l	72.1	0.789	1.3727	208.1	337.2
Pentanal	$CH_3(CH_2)_3CHO$	l	86.1	0.809	1.3944	181.6	376.1
Propenal	CH_2CHCHO	l	56.1	0.841	1.4017	186.1	325.1
Benzaldehyde	C_6H_5CHO	l	106.1	1.050	1.5463	247.1	451.1
KETONES							
Propanone (acetone)	CH_3COCH_3	l	58.1	0.789	1.3587	177.8	329.3
Butanone	$CH_3CH_2COCH_3$	l	72.1	0.805	1.3788	186.8	352.7
Pentan-2-one	$CH_3CH_2CH_2COCH_3$	l	86.1	0.814	1.3923	195.3	375.1
Pentan-3-one	$CH_3CH_2COCH_2CH_3$	l	86.1	0.814	1.3924	233.3	374.8
3-Methylbutanone	$(CH_3)_2CHCOCH_3$	l	86.1	0.805	1.3880	181.1	367.1
Hexan-2-one	$CH_3(CH_2)_3COCH_3$	l	100.2	0.811	1.4007	216.1	401.1
Cyclohexanone	$CH_2(CH_2)_4CO$	l	98.1	0.948	1.4507	256.7	428.7
Phenylethanone	$C_6H_5COCH_3$	l	120.2	1.028	1.5342	293.6	475.7
CARBOXYLIC ACIDS							
Methanoic (formic)	HCO_2H	l	46.0	1.220	1.3714	281.5	373.7
Ethanoic (acetic)	CH_3CO_2H	l	60.1	1.049	1.3719	289.7	391.0
Propanoic	$CH_3CH_2CO_2H$	l	74.1	0.993	1.3865	252.3	414.1
Butanoic	$CH_3CH_2CH_2CO_2H$	l	88.1	0.958	1.3980	268.6	438.6
2-Methylpropanoic	$(CH_3)_2CHCO_2H$	l	88.1	0.950	1.3930	227.0	426.3
Chloroethanoic	$ClCH_2CO_2H$	s	94.5	1.404	1.4351	336.1	460.9
Dichloroethanoic	Cl_2CHCO_2H	l	128.9	1.563	—	286.6	467.1
Trichloroethanoic	Cl_2CCO_2H	s	163.4	1.617	—	331.1	470.6
1-Aminoethanoic (glycine)	$NH_2CH_2CO_2H$	s	75.1	1.607	—	535[dec]	—

[dec] Decomposes.

Compound	$\Delta H_c^{\ominus}$ kJ mol^{-1}	$\Delta H_f^{\ominus}$ kJ mol^{-1}	$\Delta G_f^{\ominus}$ kJ mol^{-1}	$S^{\ominus}$ J mol^{-1} K^{-1}	$\frac{p}{D}$	ε_r	Notes
ALCOHOLS (continued)							
Butan-1-ol	−2675.6	−327.4	−168.9	228.0	1.66	—	**Pv**(100)
Pentan-1-ol	−3328.7	−353.6	−163.3	259.0	—	13.9	—
Hexan-1-ol	−3983.8	−377.8	−160.0	289.5	1.60	13.3	—
Heptan-1-ol	−4637.6	−403.3	−150.0	325.9	1.71	—	—
Octan-1-ol	−5293.6	−426.6	−136.4	354.4	1.68	10.3	—
Ethane-1,2-diol	−1179.5	−454.8	−323.2	166.9	2.00†	37.7	—
Propane-1,2,3-triol	−1655.2	−668.5	—	—	—	42.5	—
2-Methylpropan-2-ol	−2643.8	−359.2	—	—	—	—	—
Cyclohexanol	−3727.0	−348.8	−134.2	—	—	15.0	**Pv**(100)
ETHERS							
Methoxymethane	−1460.4	−184.0	−114.1	266.7	1.32	5.02	F
Ethoxyethane	−2724.0	−279.0	−122.7	251.9	1.14	4.34	F
Methoxybenzene	−3782.9	−114.8	—	—	1.38	—	—
ALDEHYDES							
Methanal	−570.6	−108.7	−113.0	218.7	2.27†	—	**Pg**(5)
Ethanal	−1167.1	−191.5	−128.2	160.2	2.49†	21.8^E	F
Propanal	−1820.8	−217.1	−142.1	—	2.54†	18.5^F	—
Butanal	−2476.0	−241.2	−306.4	—	2.57†	13.4	—
2-Methylpropanal	−2468.3	−248.9	—	—	2.58	—	—
Pentanal	−3166.0	−230.5^{gas}	—	—	—	—	—
Propenal	—	—	—	—	—	—	—
Benzaldehyde	−3525.1	−86.8	—	—	2.96	17.4	—
KETONES							
Propanone	−1816.5	−248.0	−154.8	—	2.95	20.7	—
Butanone	−2441.5	−275.7	−156.0	—	—	18.5	—
Pentan-2-one	−3099.1	−297.4	—	—	—	15.5	—
Pentan-3-one	−3099.5	−297.0	—	—	—	—	—
3-Methylbutanone	−3097.0	−299.5	—	—	—	—	—
Hexan-2-one	−3753.8	−322.0	—	—	—	—	—
Cyclohexanone	−3519.3	−270.7	—	—	—	—	—
Phenylethanone	−4148.7	−142.5	—	—	2.96	17.4	—
CARBOXYLIC ACIDS							
Methanoic	−254.3	−425.0	−361.4	129.0	1.52	58.5^G	C **Pv**(10)
Ethanoic	−874.1	−484.5	−389.9	159.8	1.74	6.2	F
Propanoic	−1527.2	−510.7	−383.5	—	1.74	3.3^H	—
Butanoic	−2183.3	−533.9	—	—	—	2.97	—
2-Methylpropanoic	−2343.9	−373.3	—	—	—	2.71^H	—
Chloroethanoic	−715.5	—	—	—	—	12.3^J	C
Dichloroethanoic	—	—	—	—	—	8.2	C
Trichloroethanoic	−388.3	−513.8	—	—	—	4.6^J	C
1-Aminoethanoic	−981.1	−528.6	—	—	—	28.1	—

^{E} At 283 K, 400 MHz. ^{F} At 290 K, 400 MHz. ^{G} At 400 MHz. ^{H} At 283 K. ^{J} At 333 K. † Uncertain.

Compound	Formula	St	$\dfrac{M}{\text{g mol}^{-1}}$	$\dfrac{\rho}{\text{g cm}^{-3}}$	n	$\dfrac{T_m}{\text{K}}$	$\dfrac{T_b}{\text{K}}$
CARBOXYLIC ACIDS (continued)							
2-Hydroxypropanoic (lactic)	$CH_3CHOHCO_2H$	l	90.1	1.206	—	326.1	376.1
Ethanedioic (oxalic)	CO_2HCO_2H	s	90.0	1.653	—	—	430^{sub}
Hexanedioic (adipic)	$CO_2H(CH_2)_4CO_2H$	s	146.1	1.360	—	426.0	dec
Benzenesulphonic	$C_6H_5SO_3H$	s	158.2	—	—	338.1^{anh}	—
Benzoic (benzenecarboxylic)	$C_6H_5CO_2H$	s	122.1	1.266^c	—	395.3	522.0
Benzene-1,4-dicarboxylic	$C_6H_4(CO_2H)_2$	s	166.1	—	—	sub	$>573^{sub\,\dagger}$
CARBOXYLIC ACID DERIVATIVES							
Ethanoyl chloride	CH_3COCl	l	78.5	1.104	1.3898	161.1	324.0
Ethanoyl bromide	CH_3COBr	l	123.0	1.663	1.4538	175.1	349.1
Ethanoyl iodide	CH_3COI	l	170.0	1.980	1.5491	—	381.1
Ethanamide (acetamide)	CH_3CONH_2	s	59.1	1.159	1.4278	355.4	494.3
Phenylethanamide	$CH_3CONHC_6H_5$	s	135.2	1.211	—	387.4	577.1
Ethanoic anhydride	$(CH_3CO)_2O$	l	102.1	1.082	1.3901	200.0	412.7
Benzene-1,2-dicarboxylic (phthalic) anhydride	$1,2\text{-}C_6H_4(CO)_2O$	s	148.1	—	—	404.7	569.1^{sub}
ESTERS							
Methyl methanoate	HCO_2CH_3	l	60.1	0.974	1.3433	174.1	304.6
Methyl ethanoate	$CH_3CO_2CH_3$	l	74.1	0.972	1.3614	175.1	330.1
Methyl propanoate	$CH_3CH_2CO_2CH_3$	l	88.1	0.915	1.3775	185.6	353.0
Ethyl methanoate	$HCO_2CH_2CH_3$	l	74.1	0.917	1.360	192.6	327.6
Ethyl ethanoate	$CH_3CO_2CH_2CH_3$	l	88.1	0.900	1.3723	189.6	350.2
Ethyl propanoate	$CH_3CH_2CO_2CH_2CH_3$	l	102.1	0.890	1.3793	199.20	372.2
Ethyl 3-oxo-butanoate	$CH_3COCH_2CO_2C_2H_5$	l	130.1	1.028	1.4194	$<193.0^\dagger$	453.5
NITRILES							
Ethanenitrile	CH_3CN	l	41.1	0.786	1.3442	227.4	354.7
Propanenitrile	CH_3CH_2CN	l	55.1	0.782	1.3655	180.2	370.4
Butanenitrile	$CH_3(CH_2)_2CN$	l	69.1	0.791	1.384	—	372.6
Propenenitrile	$CH_2{=}CHCN$	l	53.1	0.806	1.3911	—	350.6
MISCELLANEOUS							
Carbamide (urea)	NH_2CONH_2	s	60.1	1.32	1.484	408.1	dec
Cholesterol	$C_{27}H_{44}OH$	s	386.7	1.067	—	421.6	633.1^{dec}
Cyclooctatetraene	C_8H_8	l	104.2	0.921	1.5379	268.4	413.6
1,2-Epoxyethane	C_2H_5O	l	44.1	0.882	1.3597	162.1	286.3
1,2-Epoxypropane	CH_3CHCH_2O	l	58.1	0.859	1.3670	—	307.4
Furan	$(CH)_4O$	l	68.1	0.951	1.4214	187.5	304.5
Nitrobenzene	$C_6H_5NO_2$	l	123.1	1.203	1.5523	278.8	483.9
Phenol	C_6H_5OH	s	94.1	1.076	1.5521	316.1	454.8
Pyridine	$(CH)_5N$	l	79.1	0.983	1.5102	231.1	388.6
Thiocarbamide	NH_2CSNH_2	s	76.1	1.405	—	455.1	—
Fructose (β-d-)	$C_6H_{12}O_6$	s	180.2	1.60	—	376^{dec}	—
Glucose (α-d-)	$C_6H_{12}O_6$	s	180.2	1.562	—	423.1	—
Sucrose	$C_{12}H_{22}O_{11}$	s	342.3	1.580	1.5376	458.0	—

^{sub} Sublimes. ^{dec} Decomposes. ^{c} At 288 K. † Uncertain.

Compound	$\dfrac{\Delta H_c^{\ominus}}{\text{kJ mol}^{-1}}$	$\dfrac{\Delta H_f^{\ominus}}{\text{kJ mol}^{-1}}$	$\dfrac{\Delta G_f^{\ominus}}{\text{kJ mol}^{-1}}$	$\dfrac{S^{\ominus}}{\text{J mol}^{-1}\text{K}^{-1}}$	$\dfrac{p}{\text{D}}$	ε_r	Notes
CARBOXYLIC ACIDS (continued)							
2-Hydroxypropanoic	−1343.9	−694.0	—	—	—	22.0[K]	—
Ethanedioic	−243.3	−829.5	—	—	—	—	—
Hexanedioic	−2795.7	−994.3	—	—	—	—	—
Benzenesulphonic	—	—	—	—	—	—	—
Benzoic	−3227.0	−384.9	−245.1	—	1.71	—	—
Benzene-1,4-dicarboxylic	−3189.3	−816.1	—	—	—	—	—
CARBOXYLIC ACID DERIVATIVES							
Ethanoyl chloride	—	−272.9	−208.0	200.8	2.45	15.8	**C Ps F** dec[H2O]
Ethanoyl bromide	—	−223.5	—	—	—	—	**C Ps**
Ethanoyl iodide	—	−163.5	—	—	—	—	—
Ethanamide	−1184.6	−317.0	—	—	3.44	59.0[L]	—
Phenylethanamide	−4224.9	−209.2	—	—	—	—	—
Ethanoic anhydride	−1794.2	−637.2	—	—	2.8	20.7[M]	—
Benzene-1,2-dicarboxylic anhydride	−3259.5	−460.1	—	—	—	—	—
ESTERS							
Methyl methanoate	−972.5	−386.1	—	—	—	8.5	—
Methyl ethanoate	−1592.1	−445.8	—	—	1.72	6.68	—
Methyl propanoate	−2245.6	−471.6	—	—	—	5.5	—
Ethyl methanoate	—	−371.0	—	—	1.93	—	—
Ethyl ethanoate	−2237.9	−479.3	—	—	1.78	6.02	—
Ethyl propanoate	−2893.8	−502.7	—	—	—	—	—
Ethyl 3-oxo-butanoate	−2890.3	−506.2	—	—	—	—	—
NITRILES							
Ethanenitrile	−1247.1	+31.4	—	—	3.92	—	—
Propanenitrile	−1910.5	+15.5	—	—	4.02	—	—
Butanenitrile	−2568.5	−5.8	—	—	4.07	—	—
Propenenitrile	−1756.4	+147.2	—	—	3.87	—	—
MISCELLANEOUS							
Carbamide	−632.2	−332.9	−196.8	104.6	4.56	—	—
Cholesterol	—	—	—	—	—	—	—
Cyclooctatetraene	−4545.7	+254.5	+358.1	—	—	—	—
1,2-Epoxyethane	−1262.9	−77.6	—	—	1.90	—	—
1,2-Epoxypropane	−1917.4	−122.6	—	—	—	—	—
Furan	−2083.2	−62.4	—	—	0.66	2.95	Furfuran
Nitrobenzene	−3087.9	+12.4	+141.6	—	4.22	34.9	**Ps Pv**(1)
Phenol	−3053.4	−165.0	−47.5	—	1.45	9.78[N]	**C Ps Pv**(5)
Pyridine	−2783.2	+101.2	+181.2	—	2.20	12.3	**Pv**(10) **F**
Thiocarbamide	—	−93.0	—	—	—	—	—
Fructose	−2810.2	−1265.6	—	—	—	—	—
Glucose	−2802.5	−1273.3	—	—	—	—	—
Sucrose	−5639.7	−2226.1	—	—	—	—	—

[K] At 290 K.   [L] At 356 K, 400 MHz.   [M] At 292 K.   [N] At 333 K.

References: Landolt-Börnstein, Pedley, Thermodynamics Research Centre, Wagman, Weast.

$\Delta H_f^{\ominus}$ Standard molar enthalpy change of formation.
$\Delta G_f^{\ominus}$ Standard molar Gibbs free energy of formation.   See footnote below.
$S^{\ominus}$ Standard molar entropy.

	Ion	$\Delta H_f^{\ominus}$ kJ mol^{-1}	$\Delta G_f^{\ominus}$ kJ mol^{-1}	$S^{\ominus}$ J mol^{-1} K^{-1}		Ion	$\Delta H_f^{\ominus}$ kJ mol^{-1}	$\Delta G_f^{\ominus}$ kJ mol^{-1}	$S^{\ominus}$ J mol^{-1} K^{-1}
1	Ag^+	+105.6	+77.1	+72.7	40	ClO_3^-	−99.2	−3.3	+162.3
2	$Ag(NH_3)_2^+$	−111.3	−17.2	+245.2	41	ClO_4^-	−129.3	−8.6	+182.0
3†	Ag^{2+}	+268.6	+269.0	−88.0	42	Co^{2+}	−58.2	−54.4	−113.0
4	$Ag(CN)_2^-$	+270.3	+305.4	+192.0	43	Co^{3+}	+92.0	+134.0	−305.0
5	Al^{3+}	−531.0	−485.0	−321.7	44	Cr^{2+}	−143.5	−176.1	—
6	AlF_6^{3-}	−2522.5	−2267.6	—	45	Cr^{3+}	−232.2	−204.9	—
7	$Al(OH)_4^-$	−1490.3	−1297.9	+117.0	46	$Cr(H_2O)_6^{3+}$	−1999.1	—	—
8	AsO_2^-	−429.0	−350.0	+41.4	47	CrO_4^{2-}	−881.2	−727.8	+50.2
9	AsO_4^{3-}	−888.1	−648.5	−162.8	48	$HCrO_4^-$	−878.2	−764.8	+184.1
10	Au^+	—	+163.2	—	49	$Cr_2O_7^{2-}$	−1490.3	−1301.2	+261.9
11	Au^{3+}	—	+431.8	—	50	Cs^+	−258.3	−292.0	+133.1
12	$AuCl_4^-$	−322.2	−235.2	+266.9	51	Cu^+	+71.7	+50.0	+40.6
13	$Au(CN)_2^-$	+270.3	+305.4	+132.0	52	Cu^{2+}	+64.8	+65.5	−99.6
14	BF_4^-	−1574.9	−1486.9	+180.0	53	F^-	−332.6	−278.8	−13.8
15	BH_4^-	+48.2	+114.3	+110.5	54	Fe^{2+}	−89.1	−78.9	−137.7
16	Ba^{2+}	−537.6	−560.7	+9.6	55	Fe^{3+}	−48.5	−4.6	−315.9
17	Be^{2+}	−382.8	−379.7	−129.7	56	$Fe(CN)_6^{3-}$	+561.9	+729.3	+270.3
18	Bi^{3+}	—	+82.8	—	57	$Fe(CN)_6^{4-}$	+455.6	+694.9	+95.0
19	Br^-	−121.5	−104.0	+82.4	58	Ga^{2+}	—	−88.0	—
20	Br_3^-	−130.4	−107.1	+215.5	59	Ga^{3+}	−211.7	−159.0	−331.0
21	Br_5^-	−142.2	−103.8	+316.7	60‡	H^+	0.0	0.0	0.0
22	Br_2Cl^-	−170.3	−128.4	+188.7	61	H^-	—	−215.4	—
23	BrO^-	−94.1	−33.5	+42.0	62	Hg^{2+}	+171.1	+164.4	−32.2
24	BrO_3^-	−83.7	+1.7	+163.2	63	Hg_2^{2+}	+172.4	+153.6	+84.5
25	CO_3^{2-}	−677.1	−527.9	−56.9	64	I^-	−55.2	−51.6	+111.3
26	HCO_3^-	−692.0	−586.8	+91.2	65	I_3^-	−51.5	−51.5	+239.3
27	CN^-	+150.6	+172.4	+94.1	66	IO^-	−107.5	−38.5	−5.4
28	CNO^-	−146.0	−97.5	+106.6	67	IO_3^-	−221.3	−128.0	+118.4
29	CNS^-	+76.4	+92.7	+144.3	68	IO_4^-	−147.3	—	—
30	$HCOO^-$	−425.6	−351.0	+92.0	69	ICl^-	—	−161.1	—
31	CH_3COO^-	−486.0	−369.4	+86.6	70	I_2Cl^-	—	−132.6	—
32	$C_2O_4^{2-}$	−825.1	−674.0	+45.6	71	IBr_2^-	—	−123.0	—
33	$HC_2O_4^-$	−818.4	−698.4	+149.4	72	I_2Br^-	−128.0	−110.0	+197.5
34	Ca^{2+}	−542.8	−553.5	−53.1	73	K^+	−252.4	−283.3	+102.5
35	Cd^{2+}	−75.9	−77.6	−73.2	74	Li^+	−278.5	−293.3	+13.4
36	$Cd(NH_3)_4^{2+}$	−450.2	−226.4	+336.4	75	Mg^{2+}	−466.9	−454.8	−138.1
37	Cl^-	−167.2	−131.3	+56.5	76	Mn^{2+}	−220.7	−228.0	−73.6
38	ClO^-	−107.1	−36.8	+42.0	77	MnO_4^-	−541.4	−447.2	+191.2
39	ClO_2^-	−66.5	+17.2	+101.3	78	N_3^-	+275.1	+348.1	+107.9

† Measured in 4M $HClO_4$.

‡ Values for H^+ are zero according to the convention followed in this table.

$\Delta H_f^{\ominus}$ and $\Delta G_f^{\ominus}$ are respectively standard molar changes of enthalpy and Gibbs free energy for the following processes. For anions X^{n-}: $n/2\,H_2(g)$ + elements of $X \rightarrow nH^+(aq) + X^{n-}(aq)$.

For cations X^{n+}: $nH^+(aq)$ + elements of $X \rightarrow n/2\,H_2(g) + X^{n+}(aq)$. $S^{\ominus}$ is the standard molar entropy of $X^{n\pm}(aq) \mp n$(standard molar entropy of $H^+(aq)$). The values given are for 298 K, and chosen standard pressure and molality of 1 atm and 1 mol kg^{-1}.

Reference: Wagman.

Ion	$\Delta H_f^{\ominus}$ kJ mol^{-1}	$\Delta G_f^{\ominus}$ kJ mol^{-1}	$S^{\ominus}$ J mol^{-1} K^{-1}		Ion	$\Delta H_f^{\ominus}$ kJ mol^{-1}	$\Delta G_f^{\ominus}$ kJ mol^{-1}	$S^{\ominus}$ J mol^{-1} K^{-1}
79	NH_4^+	-132.5	-79.4	$+113.4$	109	HSO_3^-	-626.2	-527.8	$+139.7$
80	$N_2H_5^+$	-7.5	$+82.4$	$+151.0$	110	SO_4^{2-}	-909.3	-744.6	$+20.1$
81	NO_2^-	-104.6	-37.2	$+140.2$	111	HSO_4^-	-887.3	-756.0	$+131.8$
82	NO_3^-	-207.4	-111.3	$+146.4$	112	$S_2O_3^{2-}$	-652.3	-518.8	$+121.3$
83	Na^+	-240.1	-261.9	$+320.9$	113	$S_4O_6^{2-}$	-1224.2	-1030.5	$+259.4$
84	Ni^{2+}	-54.0	-45.6	-128.9	114	SbO_2^-	—	-340.2	—
85	OH^-	-230.0	-157.3	-10.8	115	Se^{2-}	$+85.0$	$+129.3$	$+174.0$
86	HO_2^-	-160.3	-67.4	$+23.8$	116	SeO_4^{2-}	-599.1	-441.4	$+54.0$
87	PO_3^-	-977.0	—	—	117	$HSeO_4^-$	-581.6	-452.3	$+149.4$
88	PO_4^{3-}	-1277.4	-1018.8	-222.0	118	Sn^{2+}	-8.8	-27.2	-17.0
89	$P_2O_7^{4-}$	-2271.1	-1919.2	-117.0	119	Sn^{4+}	$+30.5$	$+2.5$	-117.0
90	HPO_4^{2-}	-1292.1	-1089.3	-33.5	120	Sr^{2+}	-545.8	-559.4	-32.6
91	$H_2PO_4^-$	-1296.3	-1088.6	$+90.4$	121	TeO_3^{2-}	-596.6	—	—
92	HPO_3^{2-}	-969.0	-812.0	—	122	$Te(OH)_3^+$	-608.4	-496.2	$+111.7$
93	$H_2PO_3^-$	-969.4	-847.0	$+79.0$	123	Ti^{2+}	—	-337.5	—
94	$H_2PO_2^-$	-613.7	-512.0	—	124	Ti^{3+}	—	-302.0	—
95	PH_4^+	—	$+67.8$	—	125	Tl^+	$+5.4$	-32.4	$+125.5$
96	Pb^{2+}	-1.7	-24.4	$+10.5$	126	Tl^{3+}	$+196.6$	$+214.6$	-192.0
97	Pd^{2+}	$+169.5$	$+176.6$	-117.0	127	U^{2+}	—	-292.8	—
98	Pt^{2+}	—	$+185.8$	—	128	U^{3+}	-513.0	-312.0	-146.0
99	$PtCl_4^{2-}$	-503.3	-368.6	$+167.0$	129	U^{4+}	-591.2	-530.9	-410.0
100	$PtCl_6^{2-}$	-674.0	-490.0	$+220.1$	130	UO_2^{2+}	-1019.6	-953.5	-97.5
101	Rb^+	-251.2	-284.0	$+121.5$	131	VO_2^+	-649.8	-587.0	-42.3
102	S^{2-}	$+33.1$	$+85.8$	-14.6	132	VO^{2+}	-486.6	-446.4	-133.9
103	S_2^{2-}	$+30.1$	$+79.5$	$+28.5$	133	V^{3+}	—	-228.9	—
104	S_3^{2-}	$+25.9$	$+73.6$	$+66.1$	134	V^{2+}	—	-253.6	—
105	S_4^{2-}	$+23.0$	$+69.0$	$+103.3$	135	WO_4^{2-}	-1075.7	-920.0	$+63.0$
106	S_5^{2-}	$+21.3$	$+65.7$	$+140.6$	136	Zn^{2+}	-153.9	-147.0	-112.1
107	HS^-	-17.6	$+12.0$	$+62.8$	137	$Zn(OH)_4^{2-}$	—	-858.7	—
108	SO_3^{2-}	-635.5	-486.6	-29.0	138	$Zn(NH_3)_4^{2+}$	-533.5	-302.1	$+301.0$

5

ENTHALPY CHANGES OF NEUTRALIZATION | **5·7**

Reaction			$\Delta H^{\ominus}$/kJ mol^{-1}
HCl(aq)	$+$ NaOH(aq)	$\rightarrow$ NaCl(aq) $+$ H$_2$O	-57.9
HBr(aq)	$+$ NaOH(aq)	$\rightarrow$ NaBr(aq) $+$ H$_2$O	-57.6
HNO$_3$(aq)	$+$ NaOH(aq)	$\rightarrow$ NaNO$_3$(aq) $+$ H$_2$O	-57.6
CH$_3$CO$_2$H(aq)	$+$ NaOH(aq)	$\rightarrow$ CH$_3$CO$_2$Na(aq) $+$ H$_2$O	-56.1
HCl(aq)	$+$ NH$_3$(aq)	$\rightarrow$ NH$_4$Cl(aq)	-53.4
H$^+$(aq)	$+$ NH$_4$OH	$\rightarrow$ NH$_4^+$(aq) $+$ H$_2$O	-51.5
CH$_3$CO$_2$H(aq)	$+$ NH$_3$(aq)	$\rightarrow$ CH$_3$CO$_2$NH$_4$(aq)	-50.4
H$_2$S(aq)	$+$ OH$^-$	$\rightarrow$ HS$^-$(aq) $+$ H$_2$O	-32.2
$\frac{1}{2}$Cu^{2+}(aq)	$+$ OH$^-$	$\rightarrow$ $\frac{1}{2}$Cu(OH)$_2$(aq)	-30.1
$\frac{1}{2}$Mg^{2+}(aq)	$+$ OH$^-$	$\rightarrow$ $\frac{1}{2}$Mg(OH)$_2$(aq)	-4.4

$\Delta H^{\ominus}$ Standard molar enthalpy change of reaction at 298 K (standard pressure and molality of 1 atm and 1 mol kg^{-1}).

Reference: International Encyclopaedia of Chemical Science.

Z Mole ratio of solution $= n(H_2O)/n(X)$ where n is the amount of substance and X is the given formula of the solute.

ΔH_f Molar enthalpy change of formation for the process at 298 K and 1 atm: elements of $X^* + ZH_2O(l) \rightarrow X, ZH_2O$ (solution).

(* In states stable at 1 atm and 298 K; gases in hypothetical ideal states.)

$\Delta H_f/\text{kJ mol}^{-1}$

#	X	$Z=1$	2	5	10	20	50	100	200	500	1000	2000	∞
1	H_2O_2	−189.81	−190.46	−190.95	−191.08	−191.15	−191.15	—	—	—	—	—	−191.17
2	HF	—	−317.11	−318.67	−318.97	−319.10	−319.31	−319.41	−319.48	−319.75	−320.21	−321.33	−332.63
3	HCl	−121.55	−140.96	−155.77	−161.32	−163.85	−165.36	−165.92	−166.27	−166.57	−166.73	−166.85	−167.16
4	$HClO_4$	—	−106.27	−126.23	−129.79	−130.04	−129.54	−129.24	−129.08	−129.03	−129.03	−129.08	−129.33
5	HBr	−72.72	−93.72	−111.74	−116.96	−119.08	−120.16	−120.56	−120.81	−121.03	−121.16	−121.26	−121.55
6	HI	—	−34.63	−46.40	−51.61	−53.53	−54.21	−54.45	−54.60	−54.74	−54.84	−54.92	−55.19
7	HIO_3	—	—	—	—	—	—	−216.3	−216.7	−217.9	−218.5	—	—
8	SO_2	—	—	—	—	—	—	−326.58	−327.84	−329.75	−331.38	−333.22	−337.11
9	H_2SO_3	—	—	—	—	—	—	−612.41	−613.67	−615.58	−617.21	−619.01	−623.05
10	H_2SO_4	−841.79	−855.44	−871.48	−880.53	−884.92	−886.77	−887.64	−888.63	−890.49	−892.34	−894.29	−909.27
11	NH_3	−75.36	−77.66	−79.27	−79.81	−80.02	−80.15	−80.19	−80.21	−80.22	−80.21	−80.19	−79.69
12	NH_4OH	−363.49	−364.33	−365.29	−365.66	−365.86	−365.99	−366.03	−366.04	−366.05	−366.04	−366.02	−366.50
13	HNO_3	−187.63	−194.56	−202.77	−205.82	−206.75	−206.85	−206.86	−206.90	−206.97	−207.04	−207.11	−207.36
14	NH_4NO_3	—	−350.95	−347.48	−345.05	−343.10	−341.15	−340.23	−339.87	−339.67	−339.64	−339.67	−339.87
15	NH_4Cl	—	−299.63	−299.63	−299.44	−299.21	−299.09	−299.10	−299.17	−299.27	−299.35	−299.42	−299.66
16	H_3PO_4	−1274.82	−1278.63	−1283.94	−1286.50	−1287.96	−1288.95	−1289.41	−1289.83	−1290.36	−1290.90	−1291.58	−1296.71
17	$ZnCl_2$	—	—	−450.16	−456.77	−463.00	−471.54	−477.98	−481.91	−484.17	−485.43	—	−488.19
18	$ZnSO_4$	—	—	—	−1050.60	−1055.45	−1057.71	−1058.12	−1058.57	−1059.00	−1059.49	−1059.99	−1063.15
19	$Zn(NO_3)_2$	−501.2	−513.8	−543.1	−560.2	−565.93	−567.18	−567.64	−567.48	−566.97	−567.10	−567.4	−568.6
20	$CuCl_2$	—	—	—	−245.60	−253.72	−261.08	−264.55	−266.90	−269.53	−270.70	−270.79	—
21	$CuSO_4$	—	—	—	—	—	−837.99	−838.36	−838.81	−839.39	−839.98	−840.61	−844.50
22	$Cu(NO_3)_2$	—	—	—	−346.27	−350.62	−350.70	−350.41	−350.49	−351.39	—	—	−349.95
23	$MnSO_4$	—	—	—	—	−1119.55	−1122.07	−1122.69	−1123.36	−1124.28	−1124.87	−1125.79	−1130.10
24	$MgCl_2$	—	—	—	−781.65	−791.49	−795.92	−797.43	−798.35	−799.06	−799.48	−799.88	−801.15
25	$CaCl_2$	—	—	—	−862.74	−870.06	−872.91	−873.82	−874.39	−875.13	−875.54	−875.89	−877.13
26	LiOH	—	—	—	—	−505.83	−506.83	−507.28	−507.61	−507.91	−508.06	−508.18	−508.44
27	LiCl	—	—	−436.81	−441.58	−443.39	−444.35	−444.72	−444.76	−445.18	−445.29	−445.38	−445.60
28	LiBr	—	—	−392.31	−396.43	−397.97	−398.81	−399.15	−399.39	−399.60	−399.71	−399.79	−400.01
29	LiI	—	—	—	—	−331.70	−332.49	−332.82	−333.05	−333.26	−333.36	−333.44	−333.66
30	NaOH	—	—	−465.19	−469.65	−470.20	−469.83	−469.65	−469.61	−469.69	−469.76	−469.84	−470.11
31	NaF	—	—	—	—	—	−572.51	−572.31	−572.25	−572.31	−572.40	−572.48	−572.75
32	NaCl	—	—	—	−409.23	−408.42	−407.44	−407.07	−406.92	−406.91	−406.97	−407.02	−407.27
33	NaBr	—	—	—	−364.42	−363.20	−362.00	−361.87	−361.39	−361.35	−361.38	−361.43	−361.66
34	NaI	—	—	−299.03	−298.77	−297.24	−295.82	−295.30	−295.08	−295.01	−295.03	−295.08	−295.31
35	$NaNO_3$	—	—	—	−453.47	−451.41	−449.20	−448.20	−447.66	−447.36	−447.29	−447.29	−447.48
36	KOH	—	—	−475.71	−479.72	−481.19	−481.52	−481.64	−481.74	−481.89	−482.00	−482.09	−482.37
37	KF	—	—	−580.95	−583.58	−584.12	−584.25	−584.32	−584.40	−584.53	−584.63	−584.73	−585.01
38	KCl	—	—	—	—	−420.63	−419.66	−419.32	−419.19	−419.19	−419.24	−419.29	−419.53
39	KBr	—	—	—	−377.38	−375.97	−374.43	−373.92	−373.70	−373.64	−373.66	−373.70	−373.92
40	KI	—	—	—	−312.37	−310.24	−308.46	−307.80	−307.47	−307.33	−307.33	−307.35	−307.57
41	RbF	—	—	—	—	—	—	−583.21	−583.28	−583.40	−583.48	−583.56	—
42	CsCl	—	—	—	—	−427.35	−426.13	−425.61	−425.34	−425.20	−425.19	−425.22	−425.41

Reference: Wagman.

The lattice energy U of a crystal X_mY_n is the molar internal energy change for the process[A]:
$$mX^{n+}(g) + nY^{m-}(g) \rightarrow X_mY_n(s, 1\ atm)$$
The main entry in the table is the experimental value of $-U$ (based on the Born–Haber cycle and using data in this book); the second entry in parentheses is the theoretical value based on calculations.

Lattice	F^-	Cl^-	Br^- $-U/kJ\,mol^{-1}$	I^-	O^{2-}	S^{2-}
Li^+	1031 (1031)	848 (845)	803 (799)	759 (738)	2814 (2799)	2499 (2376)
Na^+	918 (912)	780 (770)	742 (735)	705 (687)	2478 (2481)	2198 (2134)
K^+	817 (807)	711 (702)	679 (674)	651 (636)	2232 (2238)	2052 (1933)
Rb^+	783 (772)	685 (677)	656 (653)	628 (617)	2161 (2163)	1944 (1904)
Cs^+	747 (739)	661 (643)	635 (623)	613 (592)	2063 (—)	1850 (—)
Ag^+	958 (920)	905 (833)	891 (816)	889 (778)	2910 (3002)	2677 (—)
Be^{2+}	3505 (3150)	3020 (3004)	2914 (2950)	2800 (2653)	4443 (4293)	3832 (3841)
Mg^{2+}	2957 (2913)	2526 (2326)	2440 (2097)	2327 (1944)	3791 (3795)	3299 (3318)
Ca^{2+}	2630 (2609)	2258 (2223)	2176 (2132)	2074 (1905)	3401 (3414)	3013 (3038)
Sr^{2+}	2492 (2476)	2156 (2127)	2075 (2008)	1963 (1937)	3223 (3217)	2848 (2874)
Ba^{2+}	2352 (2341)	2056 (2033)	1985 (1950)	1877 (1831)	3054 (3029)	2725 (2711)
Zn^{2+}	3032 (2930)	2734 (2690)	2678 (2632)	2605 (2549)	3971 (4142)	3322 (—)
Cd^{2+}	2809 (2740)	2552 (2526)	2507 (2468)	2441 (2406)	— (3806)	3121 (—)
Hg^{2+}	— (2757)	2651 (2569)	2628 (2598)	2610 (2569)	— (3907)	3037 (—)
Pb^{2+}	2522 (2460)	2269 (2229)	2219 (2169)	2163 (2086)	— (3502)	— (—)
Mn^{2+}	— (2644)	2537 (2368)	2471 (2304)	— (2212)	3745 (3724)	3238 (3376)
Cu^{2+}	— (—)	993 (904)	976 (870)	963 (833)	3189 (3273)	2865 (—)
NH_4^+	829 (834)	705 (688)	673 (658)	641 (629)	— (—)	2026 (2008)

[A] The ions on the left of the equation above are infinitely separated (zero pressure) and the temperature of the process is absolute zero. However, the difference between U and the molar internal energy change for the corresponding process at a non-zero temperature (commonly 298 K) is very small so that the corresponding value of molar *enthalpy* change at this temperature, known as lattice enthalpy, ΔH_l, is approximately related to U by
$$\Delta H_l = U - (m+n)RT$$
$$= U - (m+n) \times 2.5\,kJ\,mol^{-1}\ (at\ 298\ K)$$
ΔH_l and hence U can be obtained experimentally using the Born–Haber cycle.

Reference: Jenkins (1) and (2).

ΔU Electron affinity*, that is, molar internal energy change (at 0 K) for process: $X^{n-}(g) + e^- \rightarrow X^{(n+1)-}(g)$ (n = zero or positive).

Element	H	C	N	O	O^-	F
$\Delta U/kJ\,mol^{-1}$	$-72.774\ (\pm0.001)$	$-122.3\ (\pm0.5)$	$\leqslant 0\ (\pm19)$	$-141.1\ (\pm0.3)$	≈ 798	$-328.0\ (\pm0.3)$

Element	P	S	S^-	Cl	Br	I
$\Delta U/kJ\,mol^{-1}$	$-72\ (\pm1)$	$-200.42\ (\pm0.05)$	640	$-348.8\ (\pm0.4)$	$-324.6\ (\pm0.4)$	$-295.4\ (\pm0.4)$

* Some chemists define electron affinity for the reverse process.
Reference: Hotop.

$E^\ominus$ Standard electrode potential of aqueous system at 298 K, that is, standard e.m.f. of electrochemical cell in which Pt $(H_2(g))|2H^+(aq)$ forms the left-hand side electrode system; standard pressure and molality are chosen as 1 atm and 1 mol kg^{-1}.

$dE^\ominus/dT$ Temperature coefficient of $E^\ominus$ for 298 K.

For some of the electrode systems approximate expressions* are given for the corresponding electrode potential E, where only the lefthand side electrode system is standard. p stands for partial pressure.

	Righthand electrode system	$E^\ominus/V$	$(dE^\ominus/dT)/mV K^{-1}$
1	$[\frac{3}{2}N_2(g) + H^+(aq)]$, $[HN_3(g)]|Pt$	−3.40	−1.193
2	$Li^+(aq)|Li(s)$	−3.03	−0.534
3	$Rb^+(aq)|Rb(s)$	−2.93	−1.245
4	$K^+(aq)|K(s)$	−2.92	−1.080
5	$Ca^{2+}(aq)|Ca(s)$	−2.87	−0.175
6	$Na^+(aq)|Na(s)$	−2.71	−0.772
7	$Mg^{2+}(aq)|Mg(s)$	−2.37	+0.103
8	$Ce^{3+}(aq)|Ce(s)$	−2.33	+0.101
9	$Th^{4+}(aq)|Th(s)$	−1.90	+0.280
10	$Be^{2+}(aq)|Be(s)$	−1.85	+0.565
11	$U^{3+}(aq)|U(s)$	−1.80	−0.070
12	$Al^{3+}(aq)|Al(s)$	−1.66	+0.504
13	$Mn^{2+}(aq)|Mn(s)$	−1.19	−0.080
14	$V^{2+}(aq)|V(s)$	−1.18	
15	$[SO_4^{2-}(aq) + H_2O(l)]$, $[SO_3^{2-}(aq) + 2OH^-(aq)]|Pt$	−0.93	−1.389
16	$Zn^{2+}(aq)|Zn(s)$	−0.76	+0.091
17	$Cr^{3+}(aq)|Cr(s)$	−0.74	+0.468
18	$[As(s) + 3H^+(aq)]$, $AsH_3(g)|Pt$	−0.60	−0.050
19	$[2SO_3^{2-}(aq) + 3H_2O(l)]$, $[S_2O_3^{2-}(aq) + 6OH^-(aq)]|Pt$	−0.58	−1.146
20	$Fe(OH)_3(s)$, $[Fe(OH)_2(s) + OH^-(aq)]|Pt$	−0.56	−0.96
21	$[H_3PO_3(aq) + 2H^+(aq)]$, $[H_3PO_2(aq) + H_2O(l)]|Pt$	−0.499	−0.36

$E/V = -0.499 - 0.0591\,pH + 0.0295\,lg([H_3PO_3]/[H_3PO_2])$

22	$S(s)$, $S^{2-}(aq)|Pt$	−0.48	−0.93
23	$Fe^{2+}(aq)|Fe(s)$	−0.44	+0.052
24	$Cr^{3+}(aq)$, $Cr^{2+}(aq)|Pt$	−0.41	
25	$Cd^{2+}(aq)|Cd(s)$	−0.40	−0.093
26	$[Se(s) + 2H^+(aq)]$, $H_2Se(g)|Pt$	−0.40	−0.28
27	$Ti^{3+}(aq)$, $Ti^{2+}(aq)|Pt$	−0.37	
28	$PbSO_4(s)$, $[Pb(s) + SO_4^{2-}(aq)]|Pt$	−0.36	−1.015
29	$Co^{2+}(aq)|Co(s)$	−0.28	+0.06

$E/V = -0.277 + 0.0295\,lg([Co^{2+}]/mol\,dm^{-3})$

30	$[H_3PO_4(aq) + 2H^+(aq)]$, $[H_3PO_3(aq) + H_2O(l)]|Pt$	−0.276	−0.36

$E/V = -0.276 - 0.0591\,pH + 0.0295\,lg([H_3PO_4]/[H_3PO_3])$

31	$V^{3+}(aq)$, $V^{2+}(aq)|Pt$	−0.26	
32	$Ni^{2+}(aq)|Ni(s)$	−0.250	+0.06
33	$[2SO_4^{2-}(aq) + 4H^+(aq)]$, $[S_2O_6^{2-}(aq) + 2H_2O(l)]|Pt$	−0.22	+0.52
34	$Sn^{2+}(aq)|Sn(white, s)$	−0.14	−0.282
35	$Pb^{2+}(aq)|Pb(s)$	−0.13	−0.451

* These expressions assume that 1 mol dm^{-3} ≙ 1 mol kg^{-1} and solution ideality.

	Righthand electrode system	$E^{\ominus}/\text{V}$	$(\text{d}E^{\ominus}/\text{d}T)/\text{mV K}^{-1}$
36	$[\text{CrO}_4^{2-}(\text{aq}) + 4\text{H}_2\text{O(l)}], [\text{Cr(OH)}_3(\text{s}) + 5\text{OH}^-(\text{aq})]|\text{Pt}$	-0.13	-1.675
37	$[\text{CO}_2(\text{g}) + 2\text{H}^+(\text{aq})], [\text{CO(g)} + \text{H}_2\text{O(l)}]|\text{Pt}$	-0.10	
	$E/\text{V} = -0.103 - 0.0591\,\text{pH} + 0.0295\,\lg(p_{\text{CO}_2}/p_{\text{CO}})$		
38	$2\text{H}^+(\text{aq})|[\text{H}_2(\text{g})]\text{Pt}$	± 0	± 0
39	$[\text{HCO}_2\text{H(aq)} + \text{H}^+(\text{aq})], [\text{H}_2\text{O(l)} + \text{HCHO(aq)}]|\text{Pt}$	$+0.06$	
	$E/\text{V} = +0.056 - 0.0591\,\text{pH} + 0.0295\,\lg([\text{HCO}_2\text{H}]/[\text{HCHO}])$		
40	$\frac{1}{2}\text{S}_4\text{O}_6^{2-}(\text{aq}), \text{S}_2\text{O}_3^{2-}(\text{aq})|\text{Pt}$	$+0.09$	
41	$[2\text{H}^+(\text{aq}) + \text{S(s)}], \text{H}_2\text{S(aq)}|\text{Pt}$	$+0.14$	-0.209
	$E/\text{V} = +0.142 - 0.0591\,\text{pH} - 0.0295\,\lg([\text{H}_2\text{S}]/\text{mol dm}^{-3})$		
42	$[\text{Sn}^{4+}(\text{aq})1.0\text{M HCl}], [\text{Sn}^{2+}(\text{aq})(1.0\text{M HCl})]|\text{Pt}$	$+0.15$	
43	$\text{Cu}^{2+}(\text{aq}), \text{Cu}^+(\text{aq})|\text{Pt}$	$+0.15$	$+0.073$
44	$[4\text{H}^+(\text{aq}) + \text{SO}_4^{2-}(\text{aq})], [\text{H}_2\text{SO}_3(\text{aq}) + \text{H}_2\text{O(l)}]|\text{Pt}$	$+0.17$	$+0.81$
45	$\text{AgCl(s)}, [\text{Ag(s)} + \text{Cl}^-(\text{aq})]|\text{Pt}$	$+0.22$	-0.658
46	$[\text{PbO}_2(\text{s}) + 2\text{H}_2\text{O(l)}], [\text{Pb(OH)}_2(\text{s}) + 2\text{OH}^-(\text{aq})]|\text{Pt}$	$+0.25$	-1.194
47	$[\text{HAsO}_2(\text{aq}) + 3\text{H}^+(\text{aq})], [\text{As(s)} + 2\text{H}_2\text{O(l)}]|\text{Pt}$	$+0.25$	-0.510
48	$\text{Hg}_2\text{Cl}_2(\text{s}), [2\text{Hg(s)} + 2\text{Cl}^-(\text{aq})]|\text{Pt}$	$+0.27$	-0.317
49	$[\text{PbO}_2(\text{s}) + \text{H}_2\text{O(l)}], [\text{PbO(s)} + 2\text{OH}^-(\text{aq})]|\text{Pt}$	$+0.28$	
50	$\text{Cu}^{2+}(\text{aq})|\text{Cu(s)}$	$+0.34$	$+0.008$
51	$[\text{VO}^{2+}(\text{aq}) + 2\text{H}^+(\text{aq})], [\text{V}^{3+}(\text{aq}) + \text{H}_2\text{O(l)}]|\text{Pt}$	$+0.34$	
52	$\text{Fe(CN)}_6^{3-}(\text{aq}), \text{Fe(CN)}_6^{4-}(\text{aq})|\text{Pt}$	$+0.36$	
53	$[\text{O}_2(\text{g}) + 2\text{H}_2\text{O(l)}], 4\text{OH}^-(\text{aq})|\text{Pt}$	$+0.40$	-1.680
54	$[2\text{H}_2\text{SO}_3(\text{aq}) + 2\text{H}^+(\text{aq})], [\text{S}_2\text{O}_3^{2-}(\text{aq}) + 3\text{H}_2\text{O(l)}]|\text{Pt}$	$+0.40$	-1.26
55	$[\text{S}_2\text{O}_3^{2-}(\text{aq}) + 6\text{H}^+(\text{aq})], [2\text{S(s)} + 3\text{H}_2\text{O(l)}]|\text{Pt}$	$+0.47$	
	$E/\text{V} = +0.465 - 0.0887\,\text{pH} + 0.0148\,\lg([\text{S}_2\text{O}_3^{2-}]/\text{mol dm}^{-3})$		
56	$[\text{IO}^-(\text{aq}) + \text{H}_2\text{O(l)}], [\text{I}^-(\text{aq}) + 2\text{OH}^-(\text{aq})]|\text{Pt}$	$+0.49$	
57	$[4\text{H}_2\text{SO}_3(\text{aq}) + 4\text{H}^+(\text{aq})], [\text{S}_4\text{O}_6^{2-}(\text{aq}) + 6\text{H}_2\text{O(l)}]|\text{Pt}$	$+0.51$	-1.31
	$E/\text{V} = +0.509 - 0.0394\,\text{pH} + 0.0098\,\lg([\text{H}_2\text{SO}_3]^4/[\text{S}_4\text{O}_6^{2-}]\,\text{mcl}^3\,\text{dm}^{-9})$		
58	$\text{Cu}^+(\text{aq})|\text{Cu(s)}$	$+0.52$	-0.058
59	$[\text{TeO}_2(\text{s}) + 4\text{H}^+(\text{aq})], [\text{Te(s)} + 2\text{H}_2\text{O(l)}]|\text{Pt}$	$+0.53$	-0.370
60	$\text{I}_2(\text{aq}), 2\text{I}^-(\text{aq})|\text{Pt}$	$+0.54$	-0.148
61	$[\text{H}_3\text{AsO}_4(\text{aq}) + 2\text{H}^+(\text{aq})], [\text{HAsO}_2(\text{aq}) + 2\text{H}_2\text{O(l)}]|\text{Pt}$	$+0.56$	-0.364
62	$[\text{S}_2\text{O}_6^{2-}(\text{aq}) + 4\text{H}^+(\text{aq})], 2\text{H}_2\text{SO}_3(\text{aq})\text{h}]|\text{Pt}$	$+0.57$	$+1.10$
	$E/\text{V} = +0.57 - 0.1182\,\text{pH} + 0.0295\,\lg([\text{S}_2\text{O}_6^{2-}]/[\text{H}_2\text{SO}_3]^2\,\text{dm}^3\,\text{mol}^{-1})$		
63	$[\text{Sb}_2\text{O}_5(\text{s}) + 6\text{H}^+(\text{aq})], [2\text{SbO}^+(\text{aq}) + 3\text{H}_2\text{O(l)}]|\text{Pt}$	$+0.58$	
64	$[\text{MnO}_4^{2-}(\text{aq}) + 2\text{H}_2\text{O(l)}], [\text{MnO}_2(\text{s}) + 4\text{OH}^-(\text{aq})]|\text{Pt}$	$+0.59$	-1.778
65	$[2\text{H}^+(\text{aq}) + \text{O}_2(\text{g})], \text{H}_2\text{O}_2(\text{aq})|\text{Pt}$	$+0.68$	-1.033
	$E/\text{V} = 0.682 - 0.0591\,\text{pH} + 0.0295\,\lg(p_{\text{O}_2}/[\text{H}_2\text{O}_2]\,\text{dm}^3\,\text{mol}^{-1}\,\text{atm})$		
66	$[\text{C}_6\text{H}_4\text{O}_2(\text{aq}) + 2\text{H}^+(\text{aq})], \text{C}_6\text{H}_4(\text{OH})_2(\text{aq})|\text{Pt}$	$+0.70$	-0.731
67	$\text{Fe}^{3+}(\text{aq}), \text{Fe}^{2+}(\text{aq})|\text{Pt}$	$+0.77$	$+1.188$
68	$\frac{1}{2}\text{Hg}_2^{2+}(\text{aq})|\text{Hg(s)}$	$+0.79$	
69	$\text{Ag}^+(\text{aq})|\text{Ag(s)}$	$+0.80$	-1.000
70	$[2\text{NO}_3^-(\text{aq}) + 4\text{H}^+(\text{aq})], [\text{N}_2\text{O}_4(\text{g}) + 2\text{H}_2\text{O(l)}]|\text{Pt}$	$+0.80$	$+0.107$
71	$[\text{ClO}^-(\text{aq}) + \text{H}_2\text{O(l)}], [\text{Cl}^-(\text{aq}) + 2\text{OH}^-(\text{aq})]|\text{Pt}$	$+0.89$	-1.079
72	$2\text{Hg}^{2+}(\text{aq}), \text{Hg}_2^{2+}(\text{aq})|\text{Pt}$	$+0.92$	
73	$[\text{NO}_3^-(\text{aq}) + 3\text{H}^+(\text{aq})], [\text{HNO}_2(\text{aq}) + \text{H}_2\text{O(l)}]|\text{Pt}$	$+0.94$	-0.80
74	$[\text{HNO}_2(\text{aq}) + \text{H}^+(\text{aq})], [\text{NO(g)} + \text{H}_2\text{O(l)}]|\text{Pt}$	$+0.99$	

	Righthand electrode system	$E^{\ominus}/V$	$(dE^{\ominus}/dT)/mV\,K^{-1}$
75	$[HIO(aq) + H^+(aq)], [I^-(aq) + H_2O(l)]\|Pt$	$+0.99$	
	$E/V = +0.987 - 0.0295\,pH + 0.0295\,lg([HIO]/[I^-])$		
76	$[VO_2^+(aq) + 2H^+(aq)], [VO^{2+}(aq) + H_2O(l)]\|Pt$	$+1.00$	$+16.9$
77	$[H_6TeO_6(s) + 2H^+(aq)], [TeO_2 + 4H_2O(l)]\|Pt$	$+1.02$	$+0.13$
78	$[N_2O_4(g) + 4H^+(aq)], [2NO(g) + 2H_2O(l)]\|Pt$	$+1.03$	-0.011
79	$Br_2(l), 2Br^-(aq)\|Pt$	$+1.07$	-0.629
80	$Br_2(aq), 2Br^-(aq)\|Pt$	$+1.09$	-0.478
81	$[2IO_3^-(aq) + 12H^+(aq)], [I_2(aq) + 6H_2O(l)]\|Pt$	$+1.19$	-0.364
	$E/V = +1.19 - 0.0709\,pH + 0.0059\,lg([IO_3^-]^2/[I_2]\,mol\,dm^{-3})$		
82	$[MnO_2(s) + 4H^+(aq)], [Mn^{2+}(aq) + 2H_2O(l)]\|Pt$	$+1.23$	-0.661
83	$Tl^{3+}(aq), Tl^+(aq)\|Pt$	$+1.25$	$+0.89$
84	$[Cr_2O_7^{2-}(aq) + 14H^+(aq)], [2Cr^{3+}(aq) + 7H_2O(l)]\|Pt$	$+1.33$	-1.263
	$E/V = 1.333 - 0.1379\,pH + 0.0098\,lg([Cr_2O_7^{2-}]/[Cr^{3+}]^2\,dm^3\,mol^{-1})$		
85	$Cl_2(aq), 2Cl^-(aq)\|Pt$	$+1.36$	-1.260
86	$[PbO_2(s) + 4H^+(aq)], [Pb^{2+}(aq) + 4H_2O(l)]\|Pt$	$+1.46$	-0.238
87	$Mn^{3+}(aq), Mn^{2+}(aq)\|Pt$	$+1.49$	$+25.2$
88	$[MnO_4^-(aq) + 8H^+(aq)], [Mn^{2+}(aq) + 4H_2O(l)]\|Pt$	$+1.51$	$+0.66$
89	$2BrO_3^-(aq), [Br_2(aq) + 6H_2O(l)]\|Pt$	$+1.52$	-0.418
90	$[2HBrO(aq) + 2H^+(aq)], [Br_2(aq) + 2H_2O(l)]\|Pt$	$+1.57$	
91	$[2HClO(aq) + 2H^+(aq)], [Cl_2(aq) + 2H_2O(l)]\|Pt$	$+1.59$	
92	$[2HBrO(aq) + 2H^+(aq)], [Br_2(l) + 2H_2O(l)]\|Pt$	$+1.60$	
93	$[H_5IO_6(aq) + H^+(aq)], [IO_3^-(aq) + 3H_2O(l)]\|Pt$	$+1.60$	
94	$[2HClO(aq) + 2H^+(aq)], [Cl_2(g) + 2H_2O(l)]\|Pt$	$+1.63$	
	$E/V = +1.630 - 0.0591\,pH + 0.0295\,lg([HClO]^2/p_{Cl_2}\,mol^2\,dm^{-6}\,atm^{-1})$		
95	$[2HCl(aq) + 6H^+(aq)], [Cl_2(g) + 4H_2O(l)]\|Pt$	$+1.64$	
96	$Pb^{4+}(aq), Pb^{2+}(aq)\|Pt$	$+1.66$ (in 1.1M $HClO_4$)	
97	$[2ClO_2^-(aq) + 8H^+(aq)], [Cl_2(g) + 4H_2O(l)]\|Pt$	$+1.68$	
98	$[Cl_2O(g) + 2H^+(aq)], [Cl_2(g) + H_2O(l)]\|Pt$	$+1.68$	
99	$[PbO_2(s) + SO_4^{2-}(aq) + 4H^+(aq)], [PbSO_4(s) + 2H_2O(l)]\|Pt$	$+1.69$	$+0.326$
100	$[MnO_4^-(aq) + 4H^+(aq)], [MnO_2(s) + 2H_2O(l)]\|Pt$	$+1.70$	-0.666
	$E/V = +1.695 - 0.0788\,pH + 0.0197\,lg([MnO_4^-]/mol\,dm^{-3})$		
101	$Ce^{4+}(aq), Ce^{3+}(aq)\|Pt$	$+1.70$ (in M $HClO_4$)	
102	$[H_2O_2(aq) + 2H^+(aq)], 2H_2O(l)\|Pt$	$+1.77$	-0.658
	$E/V = 1.776 - 0.0591\,pH + 0.0295\,lg([H_2O_2]/mol\,dm^{-3})$		
103	$Co^{3+}(aq), Co^{2+}(aq)\|Pt$	$+1.81$ (in M HNO_3)	
104	$Ag^{2+}(aq), Ag^+(aq)\|Pt$	$+1.98$	
105	$S_2O_8^{2-}(aq), 2SO_4^{2-}(aq)\|Pt$	$+2.01$	-1.26
	$E/V = +2.010 + 0.0295\,lg([S_2O_8^{2-}]/[SO_4^{2-}]^2\,dm^3\,mol^{-1})$		
106	$[O_3(g) + 2H^+(aq)], [O_2(g) + H_2O(l)]/Pt$	$+2.08$	-0.483
	$E/V = +2.076 - 0.0591\,pH + 0.0295\,lg(p_{O_3}/p_{O_2})$		
107	$[F_2O(g) + 2H^+(aq)], [2F^-(aq) + H_2O(l)]\|Pt$	$+2.15$	-1.184
108	$[FeO_4^{2-}(aq) + 8H^+(aq)], [Fe^{3+}(aq) + 4H_2O(l)]\|Pt$	$+2.20$	-0.85
109	$F_2(g), 2F^-(aq)\|Pt$	$+2.87$	-1.830
110	$[H_4XeO_6(aq) + 2H^+(aq)], [XeO_3(g) + 3H_2O(l)]\|Pt$	$+3.0$	
111	$[F_2(g) + 2H^+(aq)], 2HF(aq)\|Pt$	$+3.06$	-0.60

References: Latimer, de Bethune, Parsons, US National Bureau of Standards.

The Arrhenius equation links the rate constant k with the absolute temperature T of a reaction system;

$$k = Ae^{-E_a/RT} \quad \text{or} \quad \ln(k/u) = \ln(A/u) - E_a/RT$$

where R is the gas constant, u is a unit of k, A is called the pre-exponential factor[A], and E_a the activation energy[A].

GAS PHASE REACTIONS

	Reaction	Catalyst	Rate law	Temp. range/K	u	$\ln(A/u)$	E_a/kJ mol^{-1}
1	$H_2 + I_2 \rightarrow 2HI$	None	$-d[H_2]/dt = k[H_2][I_2]$		$dm^3\,mol^{-1}\,s^{-1}$	26.30	173.2
	$2HI \rightarrow H_2 + I_2$	None	$-\tfrac{1}{2}d[HI]/dt = k[HI]^2$		$dm^3\,mol^{-1}\,s^{-1}$	24.41	182.8
		Au	$-\tfrac{1}{2}d[HI]/dt = k$			*	105.0
		Pt	$-\tfrac{1}{2}d[HI]/dt = k[HI]$			*	58.2
2	$N_2O_5 \rightarrow N_2O_4 + \tfrac{1}{2}O_2$	None	$-d[N_2O_5]/dt = k[N_2O_5]$	273–338	s^{-1}	31.53	103.4
3	$(CH_2)_3 \rightarrow CH_2{=}CH{-}CH_3$	None	$-d[(CH_2)_3]/dt = k[(CH_2)_3]$	695–810	s^{-1}	35.23	272.3
4	$CH_3CH_2Br \rightarrow CH_2{=}CH_2 + HBr$	None	$-d[C_2H_5Br]/dt = k[C_2H_5Br]$	655–705	s^{-1}	30.97	250.6
5	$CH_3CH_2Cl \rightarrow CH_2{=}CH_2 + HCl$	None	$-d[C_2H_5Cl]/dt = k[C_2H_5Cl]$	675–765	s^{-1}	33.62	254.4
6	$2NH_3 \rightarrow N_2 + 3H_2$	W	$-\tfrac{1}{2}d[NH_3]/dt = k$			*	60.2

REACTIONS IN AQUEOUS SOLUTION

	Reaction	Catalyst	Rate law	Temp. range/K	u	$\ln(A/u)$	E_a/kJ mol^{-1}
7	$5Br^- + BrO_3^- + 6H^+ \rightarrow$ $3Br_2 + 3H_2O$	None	$-d[BrO_3^-]/dt = k[Br^-][BrO_3^-][H^+]^2$ when ionic strength = 0.1 mol kg^{-1}	298–340			66.1
8	$2I^- + H_2O_2 + 2H^+ \rightarrow$ $I_2 + 2H_2O$	None	$-d[H_2O_2]/dt = k_1[I^-][H_2O_2]$ $+ k_2[I^-][H_2O_2][H^+]$	298	$dm^3\,mol^{-1}\,s^{-1}$ $dm^6\,mol^{-2}\,s^{-1}$	18.21 15.84	56.1(k_1) 43.5(k_2)
9	$HCO_2CH_3 + H_2O \rightarrow$ $HCO_2H + CH_3OH$	H^+	$-d[HCO_2CH_3]/dt = k[HCO_2CH_3][H^+]$	298	$dm^3\,mol^{-1}\,s^{-1}$	20.3	65.3
10	$HCO_2CH_3 + OH^- \rightarrow$ $HCO_2^- + CH_3OH$	None	$-d[HCO_2CH_3]/dt = k[HCO_2CH_3][OH^-]$		$dm^3\,mol^{-1}\,s^{-1}$	19.73	40.2
11	$C_{12}H_{22}O_{11} + H_2O \rightarrow$ $2C_6H_{12}O_6$ (in aqueous propanone)	H^+	$-d[C_{12}H_{22}O_{11}]/dt = k[C_{12}H_{22}O_{11}][H^+]$		$dm^3\,mol^{-1}\,s^{-1}$	34.95	107.9
12	$H_2O_2 \rightarrow H_2O + \tfrac{1}{2}O_2$	None	$-d[H_2O_2]/dt = k[H_2O_2]$				78.7
		Colloidal Pt	$-d[H_2O_2]/dt = k[H_2O_2]$			*	49.0
		Enzyme[B]	$-d[H_2O_2]/dt = \dfrac{k_1[\text{enzyme}][\text{substrate}]}{K_M + [\text{substrate}]}$			†	36.4(k_1)

[A] Values of A and E_a are subject to substantial uncertainty but, because of the necessity that the values observed should generate observed rate constants in the temperature range of interest, it is not possible to consider the uncertainties individually.

[B] K_M in the rate law is the Michaelis constant.

* Depends on the detailed nature of the catalyst surface.

† Depends on the enzyme used.

References: Bamford, US National Bureau of Standards Circular 510, Supplements Nos. 1, 2.

6

$$A_pB_q(s) \rightleftharpoons pA^+(aq) + qB^-(aq) \qquad K_{sp} = [A^+(aq)]^p[B^-(aq)]^q \qquad U\,(unit) = mol^{(p+q)}\,dm^{-3(p+q)}.$$

	Equilibrium		$\dfrac{K_{sp}\,(298\,K)}{U}$	$p+q$
1	$Al(OH)_3(s)$	$\rightleftharpoons Al^{3+}(aq) + 3OH^-(aq)$	1.0×10^{-32}	4
2	$Au(OH)_3(s)$	$\rightleftharpoons Au^{3+}(aq) + 3OH^-(aq)$	5.5×10^{-46}	4
3	$As(OH)_3(s)$	$\rightleftharpoons As^{3+}(aq) + 3OH^-(aq)$	2.0×10^{-1}	4
4	$BaSO_4(s)$	$\rightleftharpoons Ba^{2+}(aq) + SO_4^{2-}(aq)$	1.0×10^{-10}	2
5	$BaC_2O_4(s)$	$\rightleftharpoons Ba^{2+}(aq) + C_2O_4^{2-}(aq)$	1.7×10^{-7}	2
6	$BaCO_3(s)$	$\rightleftharpoons Ba^{2+}(aq) + CO_3^{2-}(aq)$	5.5×10^{-10}	2
7	$BaCrO_4(s)$	$\rightleftharpoons Ba^{2+}(aq) + CrO_4^{2-}(aq)$	1.17×10^{-10}	2
8	$Be(OH)_2(s)$	$\rightleftharpoons Be^{2+}(aq) + 2OH^-(aq)$	2.0×10^{-18} 291 K	2
9	$Bi(OH)_3(s)$	$\rightleftharpoons Bi^{3+}(aq) + 3OH^-(aq)$	4.0×10^{-31}	4
10	$CaCO_3(s)$	$\rightleftharpoons Ca^{2+}(aq) + CO_3^{2-}(aq)$	5.0×10^{-9}	2
11	$CaC_2O_4(s)$	$\rightleftharpoons Ca^{2+}(aq) + C_2O_4^{2-}(aq)$	2.3×10^{-9}	2
12	$Ca_3(PO_4)_2(s)$	$\rightleftharpoons 3Ca^{2+}(aq) + 2PO_4^{3-}(aq)$	1.0×10^{-26}	5
13	$CaC_4H_4O_6(s)$	$\rightleftharpoons Ca^{2+}(aq) + C_4H_4O_6^{2-}(aq)$	7.7×10^{-7}	2
14	$CaSO_4(s)$	$\rightleftharpoons Ca^{2+}(aq) + SO_4^{2-}(aq)$	2.0×10^{-5}	2
15	$CdCO_3(s)$	$\rightleftharpoons Cd^{2+}(aq) + CO_3^{2-}(aq)$	2.5×10^{-14}	2
16	$CdS(s)$	$\rightleftharpoons Cd^{2+}(aq) + S^{2-}(aq)$	8.0×10^{-27}	2
17	$CaF_2(s)$	$\rightleftharpoons Ca^{2+}(aq) + 2F^-(aq)$	4.0×10^{-11}	3
18	$Ce(OH)_3(s)$	$\rightleftharpoons Ce^{3+}(aq) + 3OH^-(aq)$	1.6×10^{-20}	4
19	$Cr(OH)_2(s)$	$\rightleftharpoons Cr^{2+}(aq) + 2OH^-(aq)$	1.0×10^{-17}	3
20	$Cr(OH)_3(s)$	$\rightleftharpoons Cr^{3+}(aq) + 3OH^-(aq)$	1.0×10^{-30}	4
21	$Co(OH)_2(s)$	$\rightleftharpoons Co^{2+}(aq) + 2OH^-(aq)$	6.3×10^{-16}	3
22	$Co(OH)_3(s)$	$\rightleftharpoons Co^{3+}(aq) + 3OH^-(aq)$	4.0×10^{-45}	4
23	$Cu(OH)_2(s)$	$\rightleftharpoons Cu^{2+}(aq) + 2OH^-(aq)$	2.0×10^{-19}	3
24	$Cu^{II}CrO_4(s)$	$\rightleftharpoons Cu^{2+}(aq) + CrO_4^{2-}(aq)$	3.6×10^{-6}	2
25	$Cu(IO_3)_2(s)$	$\rightleftharpoons Cu^{2+}(aq) + 2IO_3^-(aq)$	7.6×10^{-8}	3
26	$CuS(s)$	$\rightleftharpoons Cu^{2+}(aq) + S^{2-}(aq)$	6.3×10^{-36}; 1.0×10^{-41} 291 K	2
27	$CuBr(s)$	$\rightleftharpoons Cu^+(aq) + Br^-(aq)$	3.2×10^{-8}	2
28	$Cu_2S(s)$	$\rightleftharpoons 2Cu^+(aq) + S^{2-}(aq)$	2.5×10^{-48}	3
29	$Fe_4^{III}[Fe^{II}(CN)_6]_3(s)$	$\rightleftharpoons 4Fe^{3+}(aq) + 3[Fe^{II}(CN)_6]^{4-}(aq)$	1.0×10^{-51} 291 K; 3.0×10^{-41}	7
30	$FeCO_3(s)$	$\rightleftharpoons Fe^{2+}(aq) + CO_3^{2-}(aq)$	3.5×10^{-11}	2

	Equilibrium		$\dfrac{K_{sp}\,(298\,K)}{U}$	$p+q$
46	$MgCO_3(s)$	$\rightleftharpoons Mg^{2+}(aq) + CO_3^{2-}(aq)$	1.0×10^{-5}	2
47	$MgF_2(s)$	$\rightleftharpoons Mg^{2+}(aq) + 2F^-(aq)$	6.6×10^{-9}	3
48	$MgNH_4PO_4(s)$	$\rightleftharpoons Mg^{2+}(aq) + NH_4^+(aq) + PO_4^{3-}(aq)$	2.5×10^{-13}	3
49	$Hg(OH)_2(s)$	$\rightleftharpoons Hg^{2+}(aq) + 2OH^-(aq)$	6.3×10^{-24}	3
50	$HgSe(s)$	$\rightleftharpoons Hg^{2+}(aq) + Se^{2-}(aq)$	1.0×10^{-59}	2
51	$HgS(black)(s)$	$\rightleftharpoons Hg^{2+}(aq) + S^{2-}(aq)$	1.6×10^{-52}	2
52	$HgS(red)(s)$	$\rightleftharpoons Hg^{2+}(aq) + S^{2-}(aq)$	4.0×10^{-53}	2
53	$Hg_2Br_2(s)$	$\rightleftharpoons Hg_2^{2+}(aq) + 2Br^-(aq)$	5.75×10^{-23}	3
54	$Hg_2Cl_2(s)$	$\rightleftharpoons Hg_2^{2+}(aq) + 2Cl^-(aq)$	1.6×10^{-18}	3
55	$Hg_2(CN)_2(s)$	$\rightleftharpoons Hg_2^{2+}(aq) + 2CN^-(aq)$	5.0×10^{-40}	3
56	$Hg_2I_2(s)$	$\rightleftharpoons Hg_2^{2+}(aq) + 2I^-(aq)$	4.5×10^{-29}	3
57	$Hg_2C_2O_4(s)$	$\rightleftharpoons Hg_2^{2+}(aq) + C_2O_4^{2-}(aq)$	1.0×10^{-13}	2
58	$Hg_2SO_4(s)$	$\rightleftharpoons Hg_2^{2+}(aq) + SO_4^{2-}(aq)$	6.55×10^{-7}	2
59	$NiCO_3(s)$	$\rightleftharpoons Ni^{2+}(aq) + CO_3^{2-}(aq)$	6.6×10^{-9}	2
60	$Ni(OH)_2(s)$	$\rightleftharpoons Ni^{2+}(aq) + 2OH^-(aq)$	6.3×10^{-18}	3
61	$\gamma\text{-}NiS(s)$	$\rightleftharpoons Ni^{2+}(aq) + S^{2-}(aq)$	2.0×10^{-26}; 1.1×10^{-27} 291 K	2
62	$K_2PtCl_6(s)$	$\rightleftharpoons 2K^+(aq) + [PtCl_6]^{2-}(aq)$	1.1×10^{-5}	3
63	$KClO_4(s)$	$\rightleftharpoons K^+(aq) + ClO_4^-(aq)$	1.07×10^{-2}	2
64	$AgCl(s)$	$\rightleftharpoons Ag^+(aq) + Cl^-(aq)$	2.0×10^{-10}	2
65	$AgBr(s)$	$\rightleftharpoons Ag^+(aq) + Br^-(aq)$	5.0×10^{-13}	2
66	$AgI(s)$	$\rightleftharpoons Ag^+(aq) + I^-(aq)$	8.0×10^{-17}	2
67	$AgBrO_3(s)$	$\rightleftharpoons Ag^+(aq) + BrO_3^-(aq)$	6.0×10^{-10}	2
68	$AgIO_3(s)$	$\rightleftharpoons Ag^+(aq) + IO_3^-(aq)$	2.0×10^{-8}	2
69	$Ag_2CrO_4(s)$	$\rightleftharpoons 2Ag^+(aq) + CrO_4^{2-}(aq)$	3.0×10^{-12}	3
70	$AgCNS(s)$	$\rightleftharpoons Ag^+(aq) + CNS^-(aq)$	2.0×10^{-12}	2
71	$Ag_2CO_3(s)$	$\rightleftharpoons 2Ag^+(aq) + CO_3^{2-}(aq)$	6.3×10^{-12}	2
72	$Ag_2Cr_2O_7(s)$	$\rightleftharpoons 2Ag^+(aq) + Cr_2O_7^{2-}(aq)$	1.02×10^{-11}	3
73	$AgCN(s)$	$\rightleftharpoons Ag^+(aq) + CN^-(aq)$	2.3×10^{-16}	2
74	$Ag_3PO_4(s)$	$\rightleftharpoons 3Ag^+(aq) + PO_4^{3-}(aq)$	1.25×10^{-20} 291 K	4
75	$Ag_2SO_4(s)$	$\rightleftharpoons 2Ag^+(aq) + SO_4^{2-}(aq)$	1.6×10^{-5}	3

6·3 Solubility products (continued)

No.	Equilibrium	K_{sp} (298 K) / U	$p+q$
76	$Ag_2S(s) \rightleftharpoons 2Ag^+(aq) + S^{2-}(aq)$	6.3×10^{-50}	3
77	$SrCO_3(s) \rightleftharpoons Sr^{2+}(aq) + CO_3^{2-}(aq)$	1.1×10^{-10}	2
78	$SrF_2(s) \rightleftharpoons Sr^{2+}(aq) + 2F^-(aq)$	2.45×10^{-9}	3
79	$SrC_2O_4(s) \rightleftharpoons Sr^{2+}(aq) + C_2O_4^{2-}(aq)$	5.0×10^{-8}	2
80	$Sn(OH)_2(s) \rightleftharpoons Sn^{2+}(aq) + 2OH^-(aq)$	1.4×10^{-28}	3
81	$Zn(OH)_2(s) \rightleftharpoons Zn^{2+}(aq) + 2OH^-(aq)$	2.0×10^{-17}	3
82	$ZnCO_3(s) \rightleftharpoons Zn^{2+}(aq) + CO_3^{2-}(aq)$	1.4×10^{-11}	2
83	$Zn(CN)_2(s) \rightleftharpoons Zn^{2+}(aq) + 2CN^-(aq)$	2.6×10^{-13} 291 K	3
84	$ZnC_2O_4(s) \rightleftharpoons Zn^{2+}(aq) + C_2O_4^{2-}(aq)$	7.5×10^{-9}	2
85	$\alpha\text{-}ZnS(s) \rightleftharpoons Zn^{2+}(aq) + S^{2-}(aq)$	1.6×10^{-24}	2
86	$\beta\text{-}ZnS(s) \rightleftharpoons Zn^{2+}(aq) + S^{2-}(aq)$	5.0×10^{-26} 291 K	2
87	$Th(OH)_4(s) \rightleftharpoons Th^{4+}(aq) + 4OH^-(aq)$	1.25×10^{-45}	5
88	$TlCl(s) \rightleftharpoons Tl^+(aq) + Cl^-(aq)$	1.75×10^{-4}	2

No.	Equilibrium	K_{sp} (298 K) / U	$p+q$
31	$FeC_2O_4(s) \rightleftharpoons Fe^{2+}(aq) + C_2O_4^{2-}(aq)$	2.1×10^{-7}	2
32	$FeS(s) \rightleftharpoons Fe^{2+}(aq) + S^{2-}(aq)$	6.3×10^{-18}	2
33	$Fe(OH)_2(s) \rightleftharpoons Fe^{2+}(aq) + 2OH^-(aq)$	6.0×10^{-15}	3
34	$Fe(OH)_3(s) \rightleftharpoons Fe^{3+}(aq) + 3OH^-(aq)$	8.0×10^{-40}	4
35	$PbCl_2(s) \rightleftharpoons Pb^{2+}(aq) + 2Cl^-(aq)$	2.0×10^{-5}	3
36	$Pb_3(AsO_4)_2(s) \rightleftharpoons 3Pb^{2+}(aq) + 2AsO_4^{3-}(aq)$	4.1×10^{-36}	5
37	$Pb(N_3)_2(s) \rightleftharpoons Pb^{2+}(aq) + 2N_3^-(aq)$	2.6×10^{-9}	3
38	$PbBr_2(s) \rightleftharpoons Pb^{2+}(aq) + 2Br^-(aq)$	3.9×10^{-5}	3
39	$PbCO_3(s) \rightleftharpoons Pb^{2+}(aq) + CO_3^{2-}(aq)$	6.3×10^{-14}	2
40	$PbF_2(s) \rightleftharpoons Pb^{2+}(aq) + 2F^-(aq)$	2.7×10^{-8}	3
41	$PbI_2(s) \rightleftharpoons Pb^{2+}(aq) + 2I^-(aq)$	7.1×10^{-9}	3
42	$PbC_2O_4(s) \rightleftharpoons Pb^{2+}(aq) + C_2O_4^{2-}(aq)$	3.4×10^{-11}	2
43	$PbSO_4(s) \rightleftharpoons Pb^{2+}(aq) + SO_4^{2-}(aq)$	1.6×10^{-8}	2
44	$PbS(s) \rightleftharpoons Pb^{2+}(aq) + S^{2-}(aq)$	1.25×10^{-28}	2
45	$Mg(OH)_2(s) \rightleftharpoons Mg^{2+}(aq) + 2OH^-(aq)$	2.0×10^{-11}	3

6·4 Molar conductivities of ions in aqueous solution

λ_∞ Limiting molar conductivity of ions (or molar conductivity of ions at infinite dilution). Values are given for 298 K.

	Cation	$\lambda_\infty / \mathrm{S\,cm^2\,mol^{-1}}$
1	Li^+	38.7
2	Na^+	50.1
3	Ag^+	61.9
4	K^+	73.5
5	NH_4^+	73.6
6	Zn^{2+}	105.6
7	Ni^{2+}	106
8	Mg^{2+}	106.0
9	Mn^{2+}	107.0
10	Cu^{2+}	107.2
11	Fe^{2+}	108
12	Co^{2+}	110
13	Sr^{2+}	118.8
14	Pb^{2+}	118.9
15	Ca^{2+}	119.0
16	Ba^{2+}	127.2
17	Hg^{2+}	127.2
18	Al^{3+}	189
19	Fe^{3+}	205.2
20	H^+	349.6

	Anion	$\lambda_\infty / \mathrm{S\,cm^2\,mol^{-1}}$
21	IO_3^-	40.5
22	$CH_3CO_2^-$	40.9
23	HCO_3^-	44.5
24	HSO_4^-	52
25	HCO_2^-	54.6
26	F^-	55.4
27	BrO_3^-	55.8
28	MnO_4^-	61
29	ClO_3^-	64.6
30	ClO_4^-	67.3
31	NO_3^-	71.4
32	Cl^-	76.3
33	I^-	76.8
34	Br^-	78.1
35	CN^-	82
36	CO_3^{2-}	138.6
37	$C_2O_4^{2-}$	148.2
38	SO_4^{2-}	160.0
39	OH^-	199.1
40	PO_4^{3-}	240

References:
Kaye, Sillen, Weast (for Tables 6.3 and 6.4).

K_a Ionization (or dissociation) constant of an acid A (charged or otherwise) = equilibrium constant $[H^+][B^-]/[A]$ for $A \rightleftharpoons H^+ + B^-$. Likewise, K_b, the ionization (or dissociation) constant of a base B = equilibrium constant $[A^+][OH^-]/[B]$ for $B \rightleftharpoons A^+ + OH^-$.

$pK_{a\,(or\,b)} = -lg(K_{a\,(or\,b)}/mol\,dm^{-3})$. Values relate to concentrations between 0.01 and 0.1 mol dm^{-3}.

Acid or ion	Equilibrium (all in aqueous solution)	K_a(298 K)/mol dm^{-3}	pK_a(298 K)
Sulphuric	$H_2SO_4 \rightleftharpoons H^+ + HSO_4^-$	very large	
Nitric	$HNO_3 \rightleftharpoons H^+ + NO_3^-$	40	−1.4
Chromic(VI)	$H_2CrO_4 \rightleftharpoons H^+ + HCrO_4^-$	10	−1.0^U
Trichloroethanoic	$CCl_3CO_2H \rightleftharpoons H^+ + CCl_3CO_2^-$	2.3×10^{-1}	0.7^X
Iodic(V)	$HIO_3 \rightleftharpoons H^+ + IO_3^-$	1.7×10^{-1}	0.8
Dichloroethanoic	$CHCl_2CO_2H \rightleftharpoons H^+ + CHCl_2CO_2^-$	5.0×10^{-2}	1.3^X
Sulphurous	$H_2SO_3 \rightleftharpoons H^+ + HSO_3^-$	1.5×10^{-2}	1.8^V
Phosphonic	$H_3PO_3 \rightleftharpoons H^+ + H_2PO_3^-$	1.6×10^{-2}	1.8^Y
Chloric(III)	$HClO_2 \rightleftharpoons H^+ + ClO_2^-$	1.0×10^{-2}	2.0
Hydrogensulphate ion	$HSO_4^- \rightleftharpoons H^+ + SO_4^{2-}$	1.0×10^{-2}	2.0
Phosphoric(V)	$H_3PO_4 \rightleftharpoons H^+ + H_2PO_4^-$	7.9×10^{-3}	2.1^Z
Iron(III) ion	$Fe(H_2O)_6^{3+} \rightleftharpoons H^+ + Fe(H_2O)_5(OH)^{2+}$.	6.0×10^{-3}	2.2
Chloroethanoic	$CH_2ClCO_2H \rightleftharpoons H^+ + CH_2ClCO_2^-$	1.3×10^{-3}	2.9^X
Hydrofluoric	$HF \rightleftharpoons H^+ + F^-$	5.6×10^{-4}	3.3†
Nitrous	$HNO_2 \rightleftharpoons H^+ + NO_2^-$	4.7×10^{-4}	3.3
Methanoic	$HCO_2H \rightleftharpoons H^+ + HCO_2^-$	1.6×10^{-4}	3.8
Benzoic	$C_6H_5CO_2H \rightleftharpoons H^+ + C_6H_5CO_2^-$	6.3×10^{-5}	4.2
Phenylammonium ion	$C_6H_5NH_3^+ \rightleftharpoons H^+ + C_6H_5NH_2$	2.0×10^{-5}	4.6
Ethanoic^A	$CH_3CO_2H \rightleftharpoons H^+ + CH_3CO_2^-$	1.7×10^{-5}	4.8^X
1-butanoic	$CH_3(CH_2)_2CO_2H \rightleftharpoons H^+ + CH_3(CH_2)_2CO_2^-$	1.5×10^{-5}	4.8
Propanoic	$CH_3CH_2CO_2H \rightleftharpoons H^+ + CH_3CH_2CO_2^-$	1.3×10^{-5}	4.9
Aluminium ion	$Al(H_2O)_6^{3+} \rightleftharpoons H^+ + Al(H_2O)_5(OH)^{2+}$	1.0×10^{-5}	5.0
Dihydrogen phosphonate ion	$H_2PO_3^- \rightleftharpoons H^+ + HPO_3^-$	6.3×10^{-7}	6.2^Y
Carbonic^D	$H_2O + CO_2^B \rightleftharpoons H^+ + HCO_3^-$	4.5×10^{-7}	6.4^T
Hydrogenchromate(IV) ion	$HCrO_4^- \rightleftharpoons H^+ + CrO_4^{2-}$	3.2×10^{-7}	6.5^U
Hydrogen sulphide	$H_2S \rightleftharpoons H^+ + HS^-$	8.9×10^{-8}	7.1^S
Hydrogensulphite ion	$HSO_3^- \rightleftharpoons H^+ + SO_3^{2-}$	6.2×10^{-8}	7.2^V
Dihydrogenphosphate(V) ion	$H_2PO_4^- \rightleftharpoons H^+ + HPO_4^{2-}$	6.2×10^{-8}	7.2^Z
Chloric(I)	$HClO \rightleftharpoons H^+ + ClO^-$	3.7×10^{-8}	7.4
Bromic(I)	$HBrO \rightleftharpoons H^+ + BrO^-$	2.1×10^{-9}	8.7
Boric	$H_3BO_3 \rightleftharpoons H^+ + H_2BO_3^-$	5.8×10^{-10}	9.2
Ammonium ion	$NH_4^+ \rightleftharpoons H^+ + NH_3$	5.6×10^{-10}	9.3
Hydrocyanic	$HCN \rightleftharpoons H^+ + CN^-$	4.9×10^{-10}	9.3
Silicic	$H_2SiO_3 \rightleftharpoons H^+ + HSiO_3^-$	1.3×10^{-10}	9.9
Ethane-1,2-diammonium ion	$CH_2NHCH_2NH_3^+ \rightleftharpoons H^+ + CH_2NH_2CH_2NH_2$		9.9
Phenol	$C_6H_5OH \rightleftharpoons H^+ + C_6H_5O^-$	1.28×10^{-10}	9.9
Hydrogen carbonate ion	$HCO_3^- \rightleftharpoons H^+ + CO_3^{2-}$	4.8×10^{-11}	10.3^T
Butylammonium ion	$C_4H_9NH_3^+ \rightleftharpoons H^+ + C_4H_9NH_2$		10.8
Hydrogen peroxide	$H_2O_2 \rightleftharpoons H^+ + HO_2^-$	2.4×10^{-12}	11.6
Hydrogensilicate ion	$HSiO_3^- \rightleftharpoons H^+ + SiO_3^{2-}$	$1.3 \times 10^{-12†}$	11.9
Hydrogenphosphate(V) ion	$HPO_4^{2-} \rightleftharpoons H^+ + PO_4^{3-}$	4.4×10^{-13}	12.4^Z
Hydrogensulphide ion	$HS^- \rightleftharpoons H^+ + S^{2-}$	$1.2 \times 10^{-13†}$	12.9^S
Water	$H_2O \rightleftharpoons H^+ + OH^-$	$1.0 \times 10^{-14\,C}$	14.0

Base	Equilibrium (all in aqueous solution)	K_b(298 K)/mol dm^{-3}	pK_b(298 K)
Lead hydroxide	$Pb(OH)_2 \rightleftharpoons PbOH^+ + OH^-$	9.6×10^{-4}	3.0
Zinc hydroxide	$Zn(OH)_2 \rightleftharpoons ZnOH^+ + OH^-$	9.6×10^{-4}	3.0
Silver hydroxide	$AgOH \rightleftharpoons Ag^+ + OH^-$	1.1×10^{-4}	4.0
Ammonia^D	$NH_3(aq) + H_2O(l) \rightleftharpoons NH_4^+(aq) + OH^-(aq)$	1.8×10^{-5}	4.8
Hydrazine^D	$N_2H_4 + H_2O \rightleftharpoons N_2H_5^+ + OH$	1.7×10^{-7}(293 K)	5.8(293 K)
Hydroxylamine^D	$NH_2OH + H_2O \rightleftharpoons NH_3OH^+ + OH^-$	1.1×10^{-8}(293 K)	8.0(293 K)
Beryllium hydroxide	$Be(OH)_2 \rightleftharpoons Be^{2+} + 2OH^-$	5.0×10^{-11}	10.3

^{A} See also Table 6.9.
^{B} Some dissolved CO_2 forms the unionized molecule H_2CO_3 for which $K_a \approx 2 \times 10^{-4}$ mol dm^{-3} and $pK_a \approx 3.7$.
^{C} This is K_w/mol^2 dm^{-6} = $[H^+][OH^-]$/mol^2 dm^{-6}. Value is exact at 297 K. See Table 6.9.

^{D} [H$_2$O] is not included in the equilibrium constant.
^{STUVWXYZ} Compare values with the same letter. These represent successive ionizations of a parent acid.
† Discrepancy between sources.

$pK_{in} = -\lg(K_{in}/\text{mol dm}^{-3})$, where K_{in} is the indicator constant $= K_a$ for acidic indicators and K_w/K_b for basic ones (see Tables 6.5 and 6.9).

		pK_{in}(298 K)	acid	pH range	alkaline
1	Methyl violet	0.8	yellow	0.0–1.6	blue
2	Malachite green	1.0	yellow	0.2–1.8	blue/green
3	Thymol blue (acid)	1.7	red	1.2–2.8	yellow
4	Methyl yellow (in ethanol)	3.5	red	2.9–4.0	yellow
5	Methyl orange–xylene cyanole soln.	3.7	purple	3.2–4.2	green
6	Methyl orange	3.7	red	3.2–4.4	yellow
7	Bromophenol blue	4.0	yellow	2.8–4.6	blue
8	Congo red	4.0	violet	3.0–5.0	red
9	Bromocresol green	4.7	yellow	3.8–5.4	blue
10	Methyl red	5.1	red	4.2–6.3	yellow
11	Azolitmin (litmus)		red	5.0–8.0	blue
12	Bromocresol purple	6.3	yellow	5.2–6.8	purple
13	Bromothymol blue	7.0	yellow	6.0–7.6	blue
14	Phenol red	7.9	yellow	6.8–8.4	red
15	Thymol blue (base)	8.9	yellow	8.0–9.6	blue
16	Phenolphthalein (in ethanol)	9.3	colourless	8.2–10.0	red
17	Thymolphthalein	9.7	colourless	8.3–10.6	blue
18	Alizarin yellow R	12.5	yellow	10.1–13.0	orange/red

Note Most indicators are 0.1% solutions in H_2O unless stated otherwise.
Warning Certain indicators are poisonous and should be handled carefully, particularly when concentrated.

BUFFER SOLUTIONS | **6·7**

The following mixtures give the indicated pH at 298 K.

pH	x	Composition of solutions
1.0	67.0	25 cm³ of 0.2 mol dm⁻³ KCl + x cm³ of 0.2 mol dm⁻³ HCl
1.5	20.7	
2.0	6.5	
2.5	38.8	50 cm³ of 0.1 mol dm⁻³ potassium hydrogen phthalate($KHC_8O_4H_4$) + x cm³ of 0.1 mol dm⁻³ HCl
3.0	22.3	
3.5	8.2	
4.0	0.1	
4.5	8.7	50 cm³ of 0.1 mol dm⁻³ potassium hydrogen phthalate($KHC_8O_4H_4$) + x cm³ of 0.1 mol dm⁻³ NaOH
5.0	22.6	
5.5	36.6	
6.0	5.6	50 cm³ of 0.1 mol dm⁻³ potassium dihydrogen phosphate(KH_2PO_4) + x cm³ of 0.1 mol dm⁻³ NaOH
6.5	13.9	
7.0	29.1	
7.5	41.1	
8.0	46.7	

pH	x	Composition of solutions
8.5	15.2	50 cm³ of 0.025 mol dm⁻³ borax($Na_2B_4O_7.10H_2O$) + x cm³ of 0.1 mol dm⁻³ HCl
9.0	4.6	
9.5	8.8	50 cm³ of 0.025 mol dm⁻³ borax($Na_2B_4O_7.10H_2O$) + x cm³ of 0.1 mol dm⁻³ NaOH
10.0	18.3	
10.5	22.7	
11.0	4.1	50 cm³ of 0.05 mol dm⁻³ disodium hydrogen phosphate(Na_2HPO_4) + x cm³ of 0.1 mol dm⁻³ NaOH
11.5	11.1	
12.0	26.9	
12.5	20.4	25 cm³ of 0.2 mol dm⁻³ KCl + x cm³ of 0.2 mol dm⁻³ NaOH
13.0	66.0	

Reference for Tables 6.5, 6.6, and 6.7: Weast.

T Temperature. K_p Equilibrium constant in terms of partial pressures.
u Unit. See particular reaction (it may be non-dimensional). Chosen standard pressure* = 1 atm.
$\Delta H^{\ominus}$ Standard molar enthalpy change for reaction.
$\Delta G^{\ominus}$ Standard molar Gibbs free energy change for reaction.

T/K	$10^3\,K/T$	K_p/u	$\lg(K_p/u)$	$\Delta H^{\ominus}/kJ\,mol^{-1}$	$\Delta G^{\ominus}/kJ\,mol^{-1}$
REACTION	$N_2O_4(g) \rightleftharpoons 2NO_2(g)$	**u = atm**			
298	3.36	1.15×10^{-1}	-0.94	58.0	5.4
350	2.86	3.89	$+0.59$	57.9	-3.9
400	2.50	4.79×10^1	$+1.68$	57.7	-12.9
450	2.22	3.47×10^2	$+2.54$	57.6	-21.9
500	2.00	1.70×10^3	$+3.23$	57.4	-30.9
550	1.82	6.03×10^3	$+3.78$	57.2	-39.9
600	1.67	1.78×10^4	$+4.25$	57.1	-48.8
REACTION	$N_2(g) + 3H_2(g) \rightleftharpoons 2NH_3(g)$	**u = atm^{-2}**			
298	3.36	6.76×10^5	$+5.83$	-92.4	-33.3
400	2.50	4.07×10^1	$+1.61$	-96.9	-12.3
500	2.00	3.55×10^{-2}	-1.45	-101.3	13.9
600	1.67	1.66×10^{-3}	-2.78	-105.8	31.9
700	1.43	7.76×10^{-5}	-4.11	-110.2	55.1
800	1.25	6.92×10^{-6}	-5.16	-114.6	79.1
900	1.11	1.00×10^{-6}	-6.00	-119.0	103.3
1100	0.91	5.00×10^{-8}	-7.70		
REACTION	$H_2(g) + CO_2(g) \rightleftharpoons H_2O(g) + CO(g)$	**u = 1**			
298	3.36	1.00×10^{-5}	-5.00	41.2	28.5
500	2.00	7.76×10^{-3}	-2.11	40.5	20.2
700	1.43	1.23×10^{-1}	-0.91	39.9	12.2
800	1.25	2.88×10^{-1}	-0.54	39.5	8.2
900	1.11	6.03×10^{-1}	-0.22	39.1	4.2
1000	1.00	9.55×10^{-1}	-0.02	38.8	0.3
1100	0.91	1.45	$+0.16$	38.5	-3.47
1200	0.83	2.10	$+0.32$	38.1	-7.36
1300	0.77	2.82	$+0.45$	37.8	-11.1
REACTION	$2SO_2(g) + O_2(g) \rightleftharpoons 2SO_3(g)$	**u = atm^{-1}**			
298	3.36	4.0×10^{24}	$+24.60$	-197	
500	2.00	2.5×10^{10}	$+10.40$		
700	1.43	3.0×10^4	$+4.48$		
1100	0.91	1.3×10^{-1}	-0.89		
REACTION	$H_2(g) + I_2(g) \rightleftharpoons 2HI(g)$	**u = 1**			
298	3.36	794	$+2.9$	-9.6	
500	2.00	160	$+2.2$		
700	1.43	54	$+1.7$		
764	1.31	46 (experimental value)			
1100	0.91	25	$+1.4$		

T/K	$10^3\,K/T$	K_p/u	$\lg(K_p/u)$	$\Delta H^{\ominus}/kJ\,mol^{-1}$
REACTION	$N_2(g) + O_2(g) \rightleftharpoons 2NO(g)$	**u = 1**		
293	3.36	4×10^{-31}	-30.4	180
700	1.43	5×10^{-13}	-12.3	
1100	0.91	4×10^{-8}	-7.4	
1500	0.67	1×10^{-5}	-5.0	

*For reactions involving *only* gases, the choice is immaterial for $\Delta G^{\ominus}$ when u = 1 or for any $\Delta H^{\ominus}$.

T/K	$10^3\,K/T$	K_p/u	$\lg(K_p/u)$	$\Delta H^\ominus/kJ\,mol^{-1}$	$\Delta G^\ominus/kJ\,mol^{-1}$

REACTION $H_2O(g) + C(s) \rightleftharpoons H_2(g) + CO(g)$ $u = atm$

298	3.36	1.00×10^{-16}	-16.0	131.3	91.3
500	2.00	2.52×10^{-7}	-6.60	134.4	63.3
700	1.43	2.82×10^{-3}	-2.55	137.6	34.2
800	1.25	5.37×10^{-2}	-1.27	139.1	19.5
900	1.11	5.75×10^{-1}	-0.24	140.7	4.18
1000	1.00	3.72	$+0.57$	142.3	-11.0
1100	0.91	1.70×10^1	$+1.23$	143.8	-26.0
1200	0.83	6.60×10^1	$+1.82$	145.4	-41.8
1300	0.77	2.04×10^2	$+2.31$	146.9	-57.6

REACTION $Ag_2CO_3(s) \rightleftharpoons Ag_2O(s) + CO_2(g)$ $u = atm$

298	3.36	3.16×10^{-6}	-5.5	81.6	31.4
350	2.86	3.98×10^{-4}	-3.4	81.2	23.0
400	2.50	1.41×10^{-2}	-1.85	80.3	14.2
450	2.22	1.86×10^{-1}	-0.73	79.9	6.3
500	2.00	1.48	$+0.17$	79.5	-1.7
550	1.82	8.91	$+0.95$	78.7	-10.0
600	1.67	63.1	$+1.8$	78.2	-20.5

REACTION $CaCO_3(s) \rightleftharpoons CaO(s) + CO_2(g)$ $u = atm$

298	3.36	1.6×10^{-23}	-22.8	177.8	130.1
500	2.00	6.3×10^{-11}	-10.2	177.4	97.5
700	1.43	1.3×10^{-5}	-4.9	177.0	65.3
800	1.25	5.0×10^{-4}	-3.3	177.0	49.8
900	1.11	1.0×10^{-2}	-2.0	177.0	33.9
1000	1.00	1.3×10^{-1}	-0.9	176.6	18.0
1100	0.91	7.9×10^{-1}	-0.1	176.6	2.1
1200	0.83	4.0	$+0.6$	176.1	-13.8
1300	0.77	15.9	$+1.2$	176.1	-29.7

6

VARIATION WITH TEMPERATURE OF WATER AND ETHANOIC ACID PROPERTIES | **6·9**

κ(water)   Electrolytic conductivity for pure water.
$K_a(CH_3CO_2H)$ Ionization constant for ethanoic acid.
K_w Ionization constant (ionic product) for water.

Temperature/K	273	283	293	298	303	313	333	373
κ(water)/$10^{-8}\,\Omega^{-1}\,cm^{-1}$	1.2	2.3	4.2	5.5	7.0	11.3	17	
$K_w/10^{-14}\,mol^2\,dm^{-6}$	0.11	0.29	0.68	1.01	1.47	2.92	5.6	51.3
$K_a(CH_3CO_2H)/10^{-5}\,mol\,dm^{-3}$	1.66	1.73	1.75	1.75	1.75	1.70	1.63	

Reference: International Encyclopaedia of Chemical Science.

This table gives values of T/K where T is the temperature at which the vapour pressure of the liquid, p, reaches the indicated value.

	p/Torr					760 Torr / p/atm 1	p/atm					
	1	10	40	100	400	760 / 1	2	5	10	20	40	60
Ammonia	164.1[s]	181.3[s]	194.0[s]	204.8	227.8	239.6	254.5	277.9	298.9	323.3	352.1	371.5
Hydrogen	9.9[s]	11.9[s]	13.6[s]	15.3	18.7	20.7	23.0	27.2	31.4			
Oxygen	54.1[s]	62.6	69.1	74.4	84.4	90.3	97.2	108.7	120.0	133.2	149.1	
CCl₂F₂	154.7	175.4	191.6	204.6	229.3	243.4	261.0	289.3	315.6	347.2	373.2	
CHCl₂F	181.9	205.7	224.4	239.3	267.0	282.1	301.6	332.2	360.2	394.4	435.8	
CCl₄	223.2[s]	253.6	277.5	296.2	331.0	349.9	375.2	414.9	451.2	495.2	549.2	
CO₂	138.9[s]	153.7[s]	164.6[s]	173.0[s]	187.5[s]	195.0[s]	204.1[s]	216.5	233.7	254.3	279.1	295.6
CS₂	199.4	228.5	250.7	268.1	301.2	319.7	342.3	378.0	409.5	448.7	496.0	529.2
Methanol	229.2	257.0	278.2	294.4	323.1	337.9	357.2	385.7	411.2	441.0	476.7	497.2
Ethanol	241.9	270.9	292.2	308.1	336.7	351.6	370.7	399.2	425.0	456.2	491.2	515.2
Propan-1-ol	258.2	287.9	309.6	326.0	355.2	371.0	390.2	422.2	450.2	484.0	523.2	
Propan-2-ol	247.1	275.6	297.0	312.7	341.0	355.7	374.5	403.4	428.9	459.2	493.4	
Butan-1-ol	272.0	303.4	326.6	343.3	374.0	390.7	413.0	445.7	476.2	510.2	550.2	
Butan-2-ol	261.0	290.1	311.3	327.3	357.1	372.7	391.4	420.7	445.2	477.2	524.2	
Benzene	236.5[s]	261.7[s]	280.8	299.3	333.8	353.3	377.0	415.7	452.0	494.7	545.5	
Methane	67.3[s]	77.7[s]	85.5[s]	91.8	104.4	111.7	120.9	134.9	148.4	164.7	186.9	
Ethane	113.7	130.3	143.4	153.9	173.5	184.6	198.2	220.4	241.2	266.8	296.8	
Propane	144.3	164.7	180.8	193.6	217.6	231.1	247.6	274.6	300.1	331.3	368.0	
Butane	171.7	195.4	214.1	229.0	256.9	272.7	292.0	323.2	352.7	389.2		
Pentane	196.6	223.1	244.0	260.6	291.7	309.3	331.2	365.6	397.9	437.5		
Hexane	219.3	248.2	270.9	289.0	322.8	341.9	366.2	404.9	439.8	482.6		
Heptane	239.2	271.1	295.5	315.0	351.2	371.6	398.0	438.9	476.0	520.7		
Octane	259.2	292.4	318.3	338.9	377.2	398.8	425.9	469.4	509.0	554.6		
Nonane	274.6	311.2	339.2	361.3	401.4	424.0						
Decane	289.7	328.9	358.7	381.8	423.8	447.3						
Ethoxyethane	198.9	225.1	245.5	261.7	291.1	307.8	329.2	363.2	395.2	432.2		

[s] Solid.

References: Weast, American Institute of Physics Handbook.

t Temperature. p Saturation vapour pressure.

ICE   **WATER**

$t/°C$	$p/10^2$ Pa	$t/°C$	$p/10^2$ Pa	$t/°C$	$p/10^2$ Pa	$t/°C$	$p/10^2$ Pa
−90	0.000093	0	6.10	50	123	92	756
−80	0.00053	5	8.72	55	157	93	785
−70	0.0026	10	12.3	60	199	94	814
−60	0.0108	15	17.0	65	250	95	845
−50	0.0394	20	23.4	70	312	96	877
−40	0.1290	25	31.7	75	385	97	909
−30	0.3810	30	42.4	80	473	98	943
−20	1.03	35	56.2	85	578	99	979
−10	2.60	40	73.8	90	701	100	1013
0	6.10	45	95.8	91	728		

Reference: Weast, American Institute of Physics Handbook.

DIPOLE MOMENTS OF INORGANIC MOLECULES | **6·12**

p Dipole moment in the vapour phase: unit chosen is the common non-SI unit the debye (D) $\cong 3.34 \times 10^{-30}$ C m.

The dipole moments of organic molecules are given in Table 5.5 'Organic compounds: physical and thermochemical data'.

	Compound	p/D		Compound	p/D		Compound	p/D
1	AsH_3	0.16	10	HI	0.42	19	$Ni(CO)_4$	0
2	CH_4	0	11	H_2O	1.84	20	NO	0.16
3	CO	0.10	12	H_2O_2	2.13	21	NO_2	0.40
4	CO_2	0	13	H_2S	0.92	22	N_2O	0.17
5	CS_2	0	14	H_2Se	0.40	23	PCl_3	0.78
6	HBr	0.80	15	H_2Te	0.20	24	PH_3	0.55
7	HCl	1.05	16	NF_3	0.22	25	SbH_3	0.12
8	HCN	2.80	17	NH_3	1.48	26	SiH_4	0
9	HF	1.91	18	N_2H_4	1.84	27	SO_2	1.63

Reference: Weast.

6

The equilibrium $ML_{n-1} + L \rightleftharpoons ML_n$, where M denotes a metal ion and L a ligand, has stepwise stability constants:
$$K_n = [ML_n]/[ML_{n-1}][L] \, dm^3 \, mol^{-1}$$

For example, $Cu(H_2O)_3Cl^+(aq) + Cl^-(aq) \rightleftharpoons Cu(H_2O)_2Cl_2(aq) + H_2O$ has the stability constant:
$K_2 = [Cu(H_2O)_2Cl_2]/[Cu(H_2O)_3Cl^+][Cl^-] \, dm^3 \, mol^{-1}$, where $\lg K_2 = 1.6$.

If a total of p ligand ions (or molecules) can form a complex, the overall stability constant is:
$$K_{(p)} = K_1 K_2 \ldots K_p = [ML_p]/[M][L]^p \, dm^{3p} \, mol^{-p}$$

Values of K_n and $K_{(p)}$ are very sensitive to temperature and to the concentration of other ions present. Values given here are for infinite dilution in pure water and may have been calculated from a formula rather than measured.

Ion	Ligand	$\lg K_1$	$\lg K_2$	$\lg K_3$	$\lg K_4$	$\lg K_5$	$\lg K_{(p)}(T)$	p	T/K
$Cu(H_2O)_4^{2+}$	Cl^-	2.80	1.60	0.49	0.73	—	5.62	4	291
	NH_3	4.25	3.61	2.98	2.24	−0.52	13.08	4	291
	$C_7H_5O_3^{-A}$	10.6	6.3	—	—	—	16.9	2	
	$C_6H_6O_2^{B}$	17.0	8.0	—	—	—	25.0	2	
Fe^{2+}	OH^-	5.7	—	—	—	—	—	—	298
Fe^{3+}	SCN^-	2.95	1.94	1.4	0.8	0.02	7.1	5	298
	F^-	5.30	4.46	3.22	2.00	0.36	15.34	5	
Co^{2+}	NH_3	1.99	1.51	0.93	0.64	0.06	4.39	6	303
Ni^{2+}	NH_3	2.67	2.12	1.61	1.07	0.63	8.01	6	303
Ag^+	NH_3	3.32	3.92	—	—	—	7.23	2	298
Zn^{2+}	NH_3	2.18	2.25	2.31	1.96	—	8.70	4	303
Cd^{2+}	NH_3	2.51	1.96	1.30	0.79	—	6.65	4	303
	CN^-	5.18	4.42	4.32	3.19	—	17.11	4	298
Hg^{2+}	NH_3^{C}	8.8	8.7	1.0	0.8	—	19.3	4	295
	CN^{-D}	18.00	16.70	3.83	2.98	—	41.52	4	293

STABILITY CONSTANTS FOR METAL ION-EDTA[E] COMPLEXES AT 293 K

In all cases these are 1:1 complexes.

Ion	$\lg K$	Ion	$\lg K$	Ion	$\lg K$	Ion	$\lg K$	Ion	$\lg K$
Ag^+	7.3	Cd^{2+}	16.6	Fe^{2+}	14.3	Mn^{2+}	14.0	Sr^{2+}	8.6
Al^{3+}	16.1	Co^{2+}	16.3	Fe^{3+}	25.1	Na^+	1.7 in 0.1M KCl	Zn^{2+}	16.5
Ba^{2+}	7.8	Co^{3+}	36 in 0.1M KCl	Hg^{2+}	21.8	Ni^{2+}	18.6		
Ca^{2+}	10.7	Cu^{2+}	18.8	Mg^{2+}	8.7	Pb^{2+}	18.0		

A 2-hydroxybenzoate

B 1,2-dihydroxybenzene

C In 2 mol dm^{-3} NH$_4$NO$_3$ solution.
D In 0.1 mol dm^{-3} NaNO$_3$ solution.

E Ethylenediamine tetra-acetic acid
(Recommended name, bis[bis(carboxymethyl)amino]ethane.)

Reference: Sillen.

Data on the properties of various materials follow. Two points should be noted.

1 The values given should be regarded in many cases as typical, rather than exact, because in most cases they depend on the composition of the sample, on its previous history, and possibly on humidity or other external factors. This is particularly true for Tables 7.3, 7.4, 7.6, 7.7, and 7.8. For some of the more variable properties, it is only possible to give a range of values.

2 It is not safe to draw conclusions from the figures given without making sure one understands the definition of the quantity measured. The figures given here may suggest why materials are useful for certain applications, but are too brief for serious design work. Only a very small sample of the more important materials is included.

Metals (See Tables 7.2–7.5.) Very few metals are used in a pure form; see the footnotes to Table 7.2. Some properties vary appreciably over the useful working range of temperature: typical variations are given in Table 7.4.

Strength and hardness may be improved by *alloying*, by *cold working* (that is, rolling, drawing, or otherwise deforming at room temperature) and by *tempering* (that is, suddenly cooling from a high temperature). Certain alloys may be hardened by sudden quenching from a high temperature, followed by reheating to a lower temperature to allow *precipitation* (or *age*) *hardening* to take place by the formation of precipitates within the metal. These processes also reduce ductility and may cause brittleness, so that usually a compromise must be achieved between strength and brittleness. Table 7.5 gives examples of the effects of these processes, which do not normally affect elastic properties or density appreciably.

When two values of σ and ε are given, the first relates to the softest annealed condition obtainable and the second to the hardest fine-grain or cold-worked condition.

Structural materials (See Table 7.6.) Many of these are of natural origin and/or of composite nature, so the variability is particularly noticeable. The theoretical strength must be divided by quite a substantial safety factor to allow for the effects of variability, shape, and joint holes. Strength may vary considerably with the orientation of the stress. The effects of moisture and fire are of great concern to the engineer.

Polymers (See Table 7.7.) The properties of commercial samples of these depend very markedly on the fillers which are used. These may be introduced to achieve a desired property or simply to save expensive material. Properties may also depend on the method of manufacture. In some cases the property of the material is far more characteristic of the filler than of the polymer, for example, fibre-glass and carbon fibre materials. In these, the function of the polymer is to stick the fibres together and transmit the stresses uniformly among them.

Glasses (See Table 7.8.) Most glasses consist predominantly of silica, SiO_2, with admixtures of other oxides which reduce the temperature required for working and impart other desirable properties, at the expense of increasing the thermal expansivity and hence the liability to crack under thermal shock.

The compositions given are typical percentages by mass and the exact values vary from batch to batch.

Ceramics (See Table 7.8.) A very large number of ceramic materials are available, some based on metal oxides rather than on clays. The four examples given are representative of the main classes of these materials. All ceramic materials are brittle and most ceramics are more or less porous.

7

Values relate to commercially pure metals at common environmental temperatures (288–298 K) and may vary between specimens. Different values may be found in other tables.

cs   Crystal system (see Tables 4.8 and 5.1.).
n Number of solid phases.
r Nearest neighbour distance.
ρ_m Density.
α Linear expansivity.

T_m Melting temperature.
Δh_S^L Specific enthalpy change (latent heat) of fusion.
c_p Specific heat capacity. $u(c_p) = 0.1\,\mathrm{J\,g^{-1}\,K^{-1}}$.
λ Thermal conductivity.

	Metal	cs	n	$\dfrac{r}{\text{nm}}$	$\dfrac{\rho_m}{\text{g cm}^{-3}}$	$\dfrac{\alpha}{10^{-6}\,\text{K}^{-1}}$	$\dfrac{T_m}{\text{K}}$	$\dfrac{\Delta h_S^L}{\text{J g}^{-1}}$	$\dfrac{c_p}{u(c_p)}$	$\dfrac{\lambda^A}{\text{W cm}^{-1}\,\text{K}^{-1}}$
1	Aluminium[B]	FCC	1	0.29	2.70	23.0	933	412	8.99	2.38
2	Antimony	RBL	3	0.29	6.68	11.0	904	163	2.07	0.18
3	Bismuth	TRG	1	0.31	9.80	13.5	544	54	0.99	0.09
4	Cadmium	HCP	1	0.30	8.64	31.5	594	54	2.31	1.00
5	Chromium	BCC	2	0.25	7.20	7.0	2130	280	4.48	0.87
6	Cobalt	FCC	3	0.25	8.90	13.7	1768	280	4.48	1.00
7	Copper[C]	FCC	1	0.26	8.92	16.7	1356	205	3.87	3.85
8	Gold	FCC	1	0.29	18.88	14.0	1337	628	1.29	3.10
9	Iridium	FCC	1	0.27	22.42	6.5	2683	138	1.30	1.48
10	Iron	BCC	4	0.25	7.86	11.7	1808	269	4.50	0.80
11	Lanthanum	HCP	4	0.38	6.14	5.0	1194	113	2.00	
12	Lead[D]	FCC	1	0.35	11.34	28.9	601	25	1.28	0.38
13	Magnesium	HCP	1	0.32	1.74	25.0	922	377	10.25	1.50
14	Manganese	CUB	4	0.89[I]	7.20	22.8	1517	262	4.77	
15	Molybdenum	BCC	1	0.27	10.2	5.0	2883		2.50	1.43
16	Nickel	FCC	2	0.25	8.90	12.8	1728	305	4.44	0.91
17	Niobium	BCC	1	0.29	8.57	7.1	2740	288	2.65	0.52
18	Platinum[E]	FCC	1	0.28	21.45	8.9	2045	113	1.33	0.73
19	Rhodium	FCC	2	0.27	12.4	8.4	2239	212	2.43	1.52
20	Silver[F]	FCC	1	0.29	10.5	19.2	1235	105	2.35	4.18
21	Sodium	BCC	1	0.37	0.97	69.6	371	113	12.22	1.35
22	Tantalum	BCC	1	0.29	16.6	6.5	3269	173	1.40	0.58
23	Tin	TET	3	0.32	7.28	21.2	505	57	2.17	0.64
24	Titanium	HCP	2	0.30	4.5	8.5	1933	322	5.22	0.20
25	Tungsten[G]	BCC	1	0.27	19.35	4.5	3683	184	1.35	1.91
26	Uranium	BCC	3	0.28	19.05	13.5	1405	650	1.16	0.28
27	Zinc[H]	HCP	1	0.27	7.14	29.7	693	100	3.86	1.13
28	Zirconium	HCP	2	0.32	6.49	5.4	2125	183	2.76	0.22

[A] These quantities vary with alloying additions and with previous treatment.
[B] Used pure for electrical conductors, air conditioning plant, and garden furniture.
[C] Used pure for electrical conductors, water pipes, and cylinders.
[D] Used pure for electrical storage batteries and cable sheaths.

[E] Used pure for electrical contacts and thermometry.
[F] Used pure for electrical contacts and decorative plating.
[G] Used pure for electric lamp filaments.
[H] Used pure for dry cells and anti-corrosion plating.
[I] Lattice constant.

ρ_e	Electrical resistivity.
ρ_e'	$= (d\rho_e/dT)/\rho_e$. Relative temperature coefficient of ρ_e.
E	Tensile (Young) modulus.
K	Bulk modulus.
G	Shear modulus.

v	Poisson ratio.
σ_u	Tensile strength.
e	Elongation at fracture.
BHN	Brinel hardness number.

		ρ_e[A]	ρ_e'[A]	E	K	G	v	σ_u[AD]	e[AE]	BHN[A]
		$10^{-8}\,\Omega\,m$	$10^{-3}\,K^{-1}$	$10^{10}\,Pa$	$10^{10}\,Pa$	$10^{10}\,Pa$		$10^7\,Pa$	%	
1	Al	2.45	4.5	7.0[B]	7.6	2.6[C]	0.34	5 to 11.4	60 to 5	20 to 27
2	Sb	39.0	5.1	7.8			0.33	1		30
3	Bi	107.0	4.6	3.2	3.2	1.2	0.30			7
4	Cd	6.8	4.2	5.0	4.2	1.9	0.21	7	50	21
5	Cr	12.7	3.0	27.9	16.0	11.5		8		70
6	Co	5.6	6.6	20.6			0.34	23 to 91		120
7	Cu	1.56	4.3	13.0[B]	13.8	4.8[C]	0.44	22 to 43	50 to 5	45 to 100
8	Au	2.04	4.0	7.8[B]	21.7	2.7[C]	0.42	12 to 22	30 to 4	33 to 58
9	Ir	4.7	4.5	51.5			0.44	22	7	170
10	Fe	8.9	6.5	21.1[B]	17.0	8.2[C]	0.29	21	50	
11	La	62.4	2.2	3.9		1.5	0.29	13	8	40
12	Pb	19.0	4.2	1.6	4.6	0.6		1.5	6	
13	Mg	3.9	4.3	4.5	3.6	1.7	0.31	9 to 22		30 to 47
14	Mn	136.0		15.7			0.39			
15	Mo	5.2	4.4	34.3				16.5		
16	Ni	6.1	6.8	20.0	17.7	7.6	0.37	34 to 99	50 to 8	90 to 210
17	Nb	15.2	2.6	10.5	17.0	3.8	0.38	26 to 69	49 to 1	80 to 160
18	Pt	9.8	3.9	16.6[B]	22.8	6.1	0.34	12		38
19	Rh	4.3	4.4	29.4			0.32	95 to 21		100
20	Ag	1.5	4.1	8.3[B]	10.4	3.0[C]		14 to 35	25 to 50	25
21	Na	4.8								
22	Ta	12.6	3.5	18.6	19.6	6.9	0.39	34 to 124	40 to 1	80 to 180
23	Sn	11.5	4.6	5.0	5.8	1.8[C]	0.33	1	60	5
24	Ti	43.1	3.8	11.6	10.8	4.4		23	54	
25	W	4.9	4.8	41.1	31.1	16.1		12	1	225
26	U	29.0		16.6		8.3	0.21	38.6	4	
27	Zn	5.5	4.2	10.8	7.2	4.3		13.9	50	
28	Zr	42.4	4.4	7.4				34 to 56	35 to 10	64 to 200

[A] These quantities vary with alloying additions and with previous treatment.
[B] Temperature coefficients $\alpha_E/10^{-4}\,K^{-1}$: Al −4.8, Cu −3.7, Au −4.8, Fe −2.3, Pt −1.0, Ag −7.5.
[C] Temperature coefficients $\alpha_G/10^{-4}\,K^{-1}$: Al −5.2, Cu −3.1, Au −3.3, Fe −2.8, Ag −4.5, Sn −5.9.

[D] σ_u is the maximum force before fracture divided by original cross-sectional area and is a measure of strength.
[E] e is the permanent fractional extension occurring before fracture and is a measure of ductility.

7

References: American Society for Metals, Copper Development Association, Hampel, Hultgren, Smithells, Woolman.

The properties of magnetic materials are greatly affected by impurities, heat treatment, and previous history of the specimen.

μ_{ri}	Initial relative permeability.
μ_{rm}	Maximum relative permeability.
B_s	$= B - \mu_0 H$, saturation magnetic flux density.
H_c	Coercivity.
B_r	Remanence.
W	Hysteresis loss per cycle per unit volume for $B_{max} = 1\,T$ (or $0.5\,T$ as indicated by superscript).
T_C	Curie temperature.
ρ_e	Electrical resistivity.
$(BH)_{max}$	Maximum energy product on hysteresis curve, figure of merit for the material.
H	Value of magnetic field strength at which $(BH)_{max}$ occurs.

SOFT MAGNETIC MATERIALS

Material	$\dfrac{\mu_{ri}}{1000}$	$\dfrac{\mu_{rm}}{1000}$	$\dfrac{B_s}{T}$	$\dfrac{H_c}{A\,m^{-1}}$	$\dfrac{B_r}{T}$	$\dfrac{W}{J\,m^{-3}}$	$\dfrac{T_C}{K}$	$\dfrac{\rho_e}{10^{-8}\,\Omega\,m}$
Fe (single crystal)		1500	2.16	12		$30^{0.5\,T}$	1043	10
Fe (99% pure)	0.25	7	2.16	80	1.3	500	1043	11
Fe (1% C steel)			2.00	600				
Fe (cast annealed)			1.70	400		1000		
Fe (2% Si dynamo)	1.0	6	2.10	60		250	1003	35
Fe (3% Si grain oriented)	5.0	40	1.98	8		30	993	48
Ni (99% pure)	0.25	2	0.61	120	0.3		631	7
Co			1.76	950			1388	9
NiFeMo(79:16:5)^A	100	1000	0.79	0.16	0.55	$0.8^{0.5\,T}$	673	60
NiFeMo(79:17:4)^B	20	100	0.87	0.16	0.5	20	733	55
NiFe(45:55)^C	2.5	2.5	1.60	24		120	673	45
NiFeCo(45:30:25)^D	0.4	2.0	1.55	96		250	988	19
NiFeCu(30:59:11)^E	0.06	0.065		240			573	70
CuMnAl(61:26:13)^F	0.8		0.48	550			603	7
MnZn(Fe$_2$O$_4$)$_2{}^G$	1.5	2.5	0.34	16	0.1	10	423	2×10^7
NiZn(Fe$_2$O$_4$)$_2{}^G$	0.8	2.5	0.37	80	0.2	14	523	10^{11}

PERMANENT MAGNETIC MATERIALS

Material	Composition	$\dfrac{B_r}{T}$	$\dfrac{H_c}{kA\,m^{-1}}$	$\dfrac{(BH)_{max}}{kJ\,m^{-3}}$ at	$\dfrac{H}{kA\,m^{-1}}$
Carbon steel	Fe(1% C)	0.9	4.4	1.6	2.7
Cobalt steel	Fe(2% Co, 4% Cr, 0.9% C)	1.0	6.0	2.9	4.2
Alnico 1	FeNiAlCo(62:21:12:5)	0.65	43	11.0	25
Magnadur	BaFe$_{12}$O$_{19}$ anisotropic	0.36	110	20	86
Ticonal C Hycomax II	FeCoNiAlTiCu(41.5:29:14:7:4.5:4)	0.85	95	32	59.5
Columax Ticonal GX	FeCoNiAlCu(51:24:14:8:3) (columnar grain orientation)	1.35	58	59.5	51
Cobalt platinum	PtCo(77:23)	0.65	360	72	206

^{A} Supermalloy. ^{B} 4-79 permalloy. ^{C} 45 permalloy. ^{E} 36 Isoperm (constant permeability alloy). ^{F} Heusler alloy.
^{D} Perminvar (constant permeability alloy). ^{G} Ferroxcube type ferrites.

References: American Institute of Physics Handbook, Kaye.

ELECTRICAL RESISTIVITY $\rho_e/10^{-8}\,\Omega\,\text{m}$

Electrical resistivity, particularly at low temperatures, is sensitive to impurity concentration and cold working.

T/K	20	40	80	160	273	373	573	973	1473
Aluminium			0.3		2.45	3.55	5.9	24.7[Liq]	32.1[Liq]
Copper	0.0008	0.058	0.29	0.77	1.55	2.38	3.61	6.7	22.3
Iron	0.007	0.37	0.64	3.55	8.70	16.61[A]	31.5[A]	85.5[A]	122.0[A]
Lead	0.59		4.7		19.8	27.8	50	107.6[Liq]	126.3[B]
Tungsten	0.005	0.066	0.60	2.33	4.82	7.3	12.4	24	39

THERMAL CONDUCTIVITY $\lambda/\text{W}\,\text{m}^{-1}\,\text{K}^{-1}$

T/K	4.2	20	76	194	273	373	573	973
Aluminium	3200	5700	420	239	238	230	226	
Copper	12000	10500	660	410	400	380	380	350
Iron	77	300	180	89	82	69	55	34
Lead	2500	59	40	36	35	34	32	
Tungsten	2600	5400	260	178	170	160	150	120

LINEAR EXPANSIVITY $\alpha/10^{-6}\,\text{K}^{-1}$

T/K	25	50	100	200	300	400	500	600	800	1200
Aluminium	0.5	3.8	12.2	20.0	23.2	24.9	26.4	28.3	33.8	
Copper	0.6	3.8	10.5	15.1	16.8	17.7	18.3	18.9	20.0	23.4
Iron		1.6	5.6	10.0	11.7	12.9	14.3	15.5	16.6	21.4[γ phase]
Lead	14.5	21.6	25.0	27.5	28.9	29.8	32.1			
Tungsten			2.7	4.1	4.5	4.6	4.6	4.7	4.8	5.1

MOLAR HEAT CAPACITY $C_p/\text{J}\,\text{mol}^{-1}\,\text{K}^{-1}$

'Classical' value is $3R = 25\,\text{J}\,\text{mol}^{-1}\,\text{K}^{-1}$ at all temperatures, where R is the gas constant.

T/K	10	20	30	50	100	200	400	800	1200
Aluminium	0.04	0.21	0.84	3.80	13.1	21.5	25.7	30.6	29.3[Liq]
Copper	0.054	0.46	1.71	6.22	16.1	22.7	25.1	27.6	30.1[Liq]
Iron	0.084	0.25	0.75	3.05	12.0	21.5	27.4	38.6	34.2[γ phase]
Lead	2.80	11.0	16.4	21.4	24.5	25.8	27.4	30.0	28.7[Liq]
Tungsten	0.046	0.326	1.42	5.93	16.0	15.0	25.0	26.5	27.8

$C_p(\text{liq Fe, 2000 K}) = 44\,\text{J}\,\text{mol}^{-1}\,\text{K}^{-1}$.

TENSILE STRENGTH $\sigma_u/10^6\,\text{Pa}$

These values should be taken to indicate the order of magnitude of changes in σ_u only. Exact values are very sensitive to impurity (alloy constituents) and heat treatment or cold working.

T/K	70	205	277	373	477	589	673	773
Aluminium	220	100	90	75	42	17		
Copper	330	260	210	190	160		130	90

[A] These figures relate to a different specimen from those at low temperatures. [B] Liq 1273 K.

References: American Society for Metals, Copper Development Association, Hampel, Smithells, Woolman.

7

R Fractional reduction in thickness by cold working.
σ_y Yield stress.
σ_p 0.1% proof stress (this is a practical reproducible measure indicating yield stress).
See Table 7.1 for explanations and definitions.

σ_u Tensile strength.
e Fractional elongation on fracture.
T Temperature of tempering.
W Mass fraction of alloying component.

COLD WORKING OF ROLLED COPPER STRIP ANNEALED INITIALLY

$R/\%$	$\sigma_p/10^8$ Pa	$\sigma_u/10^8$ Pa	$e/\%$	$R/\%$	$\sigma_p/10^8$ Pa	$\sigma_u/10^8$ Pa	$e/\%$
0	0.6	2.1	58	50	3.2	3.5	8
10	1.8	2.4	38	60	3.3	3.8	7
20	2.4	2.7	23	70	3.5	3.9	6
30	2.7	3.2	15	80	3.6	4.1	5
40	3.0	3.3	10				

TEMPERING OF PLAIN CARBON STEEL

This contains 0.45 per cent C and small additions of Si, Mn, S, and P. It is initially hardened by quenching. Tempering is carried out by maintaining at temperature T until no further change in properties is observed.

T/K	$\sigma_p/10^8$ Pa	$\sigma_u/10^8$ Pa	$e/\%$
Initial condition	7.1	9.8	12
573	6.6	9.8	15
673	6.9	9.6	17
773	6.3	9.0	21
873	5.4	7.8	25
973	4.7	6.8	28

IRON–CARBON ALLOY

Plain carbon steel containing small proportions of Si, Mn, S, and P.

$W(C)/\%$	$\sigma_y/10^8$ Pa	$\sigma_u/10^8$ Pa	$e/\%$
0.10	2.6	3.7	45
0.22	3.3	4.8	36
0.39	3.2	5.4	34
0.54	3.6	6.9	25
0.81	3.7	7.9	19
1.04	4.1	8.4	15

COPPER–ZINC ALLOY (BRASSES)

$W(Cu)/\%$	$W(Zn)/\%$	$\sigma_p/10^8$ Pa	$\sigma_u/10^8$ Pa	$e/\%$
100	0	0.45	2.3	50
90	10	0.75	2.6	63
80	20	0.90	3.0	67
70	30	1.05	3.3	70
60	40	1.35	3.6	85

References: as for Table 7.4.

Legend

- ρ Density.
- E Tensile (Young) modulus.
- σ_{max} Tensile strength.
- f Safety factor for σ_{max}. [A]
- p_{max} Compressive strength.
- T_{max} Shear strength.
- α Thermal expansivity.
- λ Thermal conductivity.
- C Combustibility.
- M.R. Moisture resistance; VG = very good. G = good, F = fair, P = poor.

Material	ρ / kg m⁻³	E / GPa	σ_{max} / MPa	f	p_{max} / MPa	T_{max} / MPa	α / μK	λ / W m⁻¹ K⁻¹	C / °C	M.R.
Metals										
Mild steel	7700	200	250[Y]	1.5	250[Y]	250[Y]	12	60	400	G[H]
High tensile steel	7700	200	340[Y]	1.3	340[Y]	330[Y]	12	60	400	VG^{pure}
Unalloyed aluminium	2700	70	60 to 120[Y]	varies	60 to 120[Y]	60 to 70[Y]	24	200	200	G
High strength Al alloy	2800	70	240 to 400[Y]	varies	240 to 400[Y]	160 to 280[Y]	24	100	200	G
Timber (dry, 18% moisture)										
Douglas fir (imported)	600	16$^{\parallel}$ / 1$^{\perp}$	18$^{\parallel}$	1.3 to 2.5	15$^{\parallel}$ / 2.5$^{\perp}$	1.9$^{\perp}$	3$^{\parallel}$ / 35$^{\perp}$	0.4$^{\parallel}$ / 0.11$^{\perp}$	burns	F[H]
Oak	720	9 av / 5 min	21$^{\parallel}$	1.3 to 2.5	15$^{\parallel}$ / 4.4$^{\perp}$	3.1$^{\perp}$	3$^{\parallel}$	0.16$^{\perp}$	burns	F[H]
Western red cedar	380	7 av / 4 min	11$^{\parallel}$	1.3 to 2.5	9$^{\parallel}$ / 1.5$^{\perp}$	1.4$^{\perp}$	3$^{\parallel}$	0.09$^{\perp}$	burns	F[H]
Balsa	200	6$^{\parallel}$ / 0.2$^{\perp}$	25$^{\parallel}$		1.5$^{\parallel}$	1.5$^{\perp}$	0.2$^{\parallel}$ / 20$^{\perp}$	0.2$^{\parallel}$ / 0.07$^{\perp}$	burns	
Boards										
Douglas fir plywood	450 to 600	7 to 11	9[C] to 15[B]	F	6[C] to 10[B]	(2)	5 to 10	0.12 to 0.19$^{\perp}$	burns	F[H]
Birch plywood	600 to 700	4 to 10	8[C] to 16[B]	F	5[C] to 8[B]	(2)	5 to 10	0.17 to 0.20$^{\perp}$	burns	F[H]
Standard hardboard[D]	800 to 1000	4 to 6	25 to 55	6	10 to 15	—		0.07 to 0.10	burns	P
Chipboard[D]	450 to 1300	2 to 4	2 to 10		5 to 10			0.10 to 0.25	burns	P
Plasterboard[D]	700 to 1300		2 to 5					0.15 to 0.20	weakens	P
Asbestos cement sheet[D]	1700 to 2000	18 to 20	20 to 35		20 to 30			0.25 to 0.35	shatters	G
Woodwool–cement[D]	400 to 800		0.3 to 2		—			0.05 to 0.10	weakens	G
Bricks and blocks										
Engineering brick	1800 to 2000	20	depends on joints	15 to 30[G]	50 to 100		6	1.00 to 1.50	does not burn	VG
Common brick	1500 to 1800	7		12[G]	7 to 70		6	0.80 to 1.20	does not burn	VG
Breeze block	1300 to 1500	20 to 40		12[G]	4 to 6		12	0.35 to 0.45	does not burn	G
Aerated concrete block[D]	600 to 1200	15 to 30			4 to 8		12	0.08 to 0.40	does not burn	G
Concrete										
1:2:4 mix	2200 to 2400	40	cracks unless reinforced	3	20 to 35		12	1.40 to 1.50	stable	VG
High strength mix	2200 to 2400	40		3	50 to 70		12	1.40 to 1.50	stable	VG
Lightweight concrete[D]	800 to 1200	15 to 40	reinforced	4	6 to 15		12	0.10 to 0.60	stable	G
Polymers										
Glass reinforced polyester[J]	1500 to 2000	5 to 7	70 to 500	4 to 10	100 to 400		20 to 30		stable[I]	FG
Rigid	1250 to 2500	2 to 4	40 to 60		40 to 60		50 to 70		200	FG
Epoxide resin	1200 to 2500	1 to 5	30 to 80		100 to 200		40 to 60		250	G

$^{\parallel}$ Parallel to grain or to plies. $^{\perp}$ Perpendicular to grain or to plies. [B] Parallel to face grain. [C] Perpendicular to face grain. [A] Maximum safe stress is σ_{max}/f. [D] E, σ_{max}, λ increase with ρ (and binder content). [F] Depends on condition of timber. [G] Includes strength of mortar. Depends on ratio of height to thickness (r). Figures given for $r = 1$. $f(12) = 2f(1)$, $f(21) = 4f(1)$.

[H] If suitably coated. [I] If special fire-resisting grade. [J] Lower figures for randomly directed fibres. Larger figures for stress parallel to unidirectional fibres. [Y] Yield strength.

7

Symbols

ρ	Density.
n	Refractive index.
σ_u	Tensile strength.
e	Elongation at fracture.
E	Tensile (Young) modulus.
τ_f	Flexural strength.
p_{max}	Compressive strength.
c_v	Specific heat capacity.
α	Thermal expansivity.
λ	Thermal conductivity.
t_m	Melting temperature.
t_g	Glass temperature.
t_{max}	Maximum service temperature.
ρ_e	Electrical resistivity.
ε_r	Relative permittivity.
p	Specific price (1982).

#	Material	ρ / g cm⁻³	n	σ_u / MPa	e / %	E / MPa	τ_f / MPa	p_{max} / MPa
1	Acrylic (e.g. Perspex[A])	1.17 to 1.20	1.49	55 to 70	4	2500 to 3500	90–130	83 to 120
2	Poly(chloroethene) (PVC)	1.25[F] 1.39[R]	1.52	20[F] 60[R]	300[F] 15–20[R]	2400 to 4100[R]	93[R]	9[F] 55[R]
3	Poly(ethene) (polythene) low density	0.92	1.51	15	600	150		
4	Poly(ethene) (polythene) high density	0.96	1.54	29	350	1000		
5	Poly(phenylethene) (polystyrene)	1.05	1.6	40	2.5	3000	60	95
6	Polyester (unsaturated) cast window filler	1.2	clear	35	1–2		120	110
7	Polyester laminate, 70% woven glass fibre	1.65	opaque	350	1–2	1100 to 1500	400	350
8	Poly(propene)	0.9	1.49	35	400	700	50	50
9	Nylon 6	1.12 to 1.14	opaque	45 to 90	100–300		35	
10	Nylon 6.6	1.1	opaque	83	60–300		42	34[B]
11	Phenolic resin (Bakelite) woodflour filler	1.35 to 1.45	brown	50	0.6	6000 to 8000	70	200
12	Neoprene[C], vulcanized	1.23 to 1.25	1.56	25	800–1000			
13	Rubber (natural) gum vulcanized	0.93	opaque	32	850			
14	Rubber (natural) carbon black filled	1.5	opaque	35	650			

#	c_v / J g⁻¹ K⁻¹	α / 10⁻⁶ K⁻¹	λ / W m⁻¹ K⁻¹	t_m / °C	t_g / °C	t_{max} / °C	ρ_e / Ωm	ε_r	p (1982) / £ kg⁻¹
1	1.5	90	0.17–0.25		90–105		$>10^{12}$	3.3	1.09
2	1 to 2[F] 0.9[R]	240[F] 50[R]	0.12–0.17		−20 to −30[F] 85[R]	70–100[F] 60[R]	10^{14}[R] 10^9–10^{12}[F]	3.5[R] 4.5[F]	0.60
3	2.3	150–200	0.33	110–130	−120	85	$>10^{14}$	2.2	0.63
4	2.3	100	0.45–0.52	120–140	−120	120	$>10^{14}$	2.35	0.66
5	1.3	70	0.04–0.14		100	65	$>10^{14}$	2.45	0.67
6	2.1	80					10^{14}	4.0	1.24
7		12					10^{13}	5.0	2.10
8	1.9	80–100	0.12	176	−10	160	$>10^{14}$	2.2	0.59
9	1.6	80	0.25	216	50	120	10^9–10^{11}	4.0	2.20
10	1.7	70	0.25	265	57	120	10^{13}	4.0	2.26
11	1.5	50	0.15			150	10^8–10^{12}	5	0.43
12	2.18		0.19		−44		10^8–10^{11}		1.65
13	2.1	220	0.13		−73	−50 to +100	10^{12}	2.5	0.55[D]
14	1.6	160	0.16			−50 to +80	10^{11}	7.0	1.00[E]

[F] Flexible PVC. [R] Rigid PVC. [A] Poly(methyl 2-methyl propenoate), or polymethyl methacrylate. [B] 1% deformation. [C] Poly(2-chlorobuta-1,3-diene), or poly(chloroprene). [D] For raw rubber. Vulcanization is part of the fabrication process, so no typical price can be given for vulcanized material. [E] Cheap general purpose availability.

Note. Carbon-fibre filled polymers are of increasing importance but the properties vary too much with fibre content to tabulate.

References: manufacturers' leaflets.

Symbols

- ρ Density.
- n Refractive index.
- V Constringence (reciprocal dispersive power)[O].
- α Thermal linear expansivity.
- T_a Annealing temperature.
- T_s Softening temperature.
- ρ_e Electrical resistivity.
- ε_r Relative permittivity (293 K, 1 MHz).
- E Tensile (Young) modulus.
- v Apparent porosity.
- σ Modulus of rupture (bending test).
- λ Thermal conductivity.
- t_f Firing temperature.

Glass	Applications	ρ / g cm^{-3}	n(589 nm)	V	α / 10^{-6} K^{-1}	σ / MPa	T_a / K	T_s / K	ρ_e / Ωm	ε_r	E / GPa	Composition (main constituents)
Vitreous silica	Tableware, immersion heaters	2.20	1.458		0.54		1413	1940	10^{12}	3.8	70	99.5% SiO$_2$
Vycor	Tableware	2.18	1.458		0.8		1183	1773	5×10^9	3.8	68	96% SiO$_2$, 3% B$_2$O$_3$
Container	Jars and bottles	2.49	1.520		8.5		821	1003	10^7	7.6	70	73% SiO$_2$, 15% Na$_2$O[D]
Sheet	Windows, electric lamps	2.46	1.510		8.5		821	1003	3×10^6	7.0	70	73% SiO$_2$, 13% Na$_2$O[E]
Borosilicate	Laboratory and ovenware	2.23	1.474		3.2		838	1093	10^8	4.6	69	80% SiO$_2$, 12% B$_2$O$_3$[F]
Alumino-silicate	Combustion tubes[A]	2.53	1.534		4.2		988	1188	3×10^{11}	6.3	89	57% SiO$_2$, 21% Al$_2$O$_3$[G]
High lead	Electrical components	4.28	1.693		9.1		703	853	6×10^{11}	9.5	53	35% SiO$_2$, 58% PbO[H]
Solder glass					8.9		613	718				16% B$_2$O$_3$, 84% PbO
Light Ba crown	Optical	2.90	1.541	59.4	8.2		843			6.90	73	57% SiO$_2$, 27% BaO[I]
Dense Ba crown	Optical	3.56	1.612	59.0	6.4		878			8.21	79	36% SiO$_2$, 45% BaO[J]
Light flint	Optical	3.26	1.578	40.8	8.0		758			6.57	60	52% SiO$_2$, 38% PbO[K]
Dense flint	Optical	3.55	1.613	36.9	8.6		733			7.42	56	48% SiO$_2$, 45% PbO[L]
Glass fibre[B]	Textiles	2.46	1.512		8.7		801	983	5×10^6	7.9	73	72% SiO$_2$, 13% Na$_2$O[M]
Glass fibre[C]	'fibreglass'	2.53	1.548		5.0		848	1103	$> 10^{15}$	6.4	77	55% SiO$_2$, 18% CaO[N]

Ceramic	ρ / g cm^{-3}	v / %	E / GPa	σ / MPa	α / 10^{-6} K^{-1}	λ / W m^{-1} K^{-1}	ρ_e / Ωm	t_f / °C
Earthenware	2.5	15 to 20	50	56	7	1.6		1400
Bone china	2.8	0 to 2	90	112	8	1.6		1500
Electrical porcelain	2.5	0	70	105	7	1.6	10^{10} to 10^{12}	1500
High alumina (90% Al$_2$O$_3$)	3.7	0	250 to 380	280 to 380	7.5	12 to 26	10^9 to 10^{12}	1950

[A] Also boiler sight glasses.   [B] Soda-lime.   [C] E glass, weather resistant.
[D] Also 10% CaO.   [E] Also 9% CaO, 3% MgO.   [F] Also 4% Na$_2$O, 2% Al$_2$O$_3$.
[G] Also 12% MgO, 6% CaO, 4% B$_2$O$_3$.   [H] Also 7% K$_2$O.   [I] Also 14% K$_2$O.
[J] Also 8% B$_2$O$_3$.   [K] Also 10% K$_2$O.   [L] Also 5% Na$_2$O.   [M] Also 9% CaO.

[N] Also 15% Al$_2$O$_3$.   [O] $V = (n_d - 1)/(n_F - n_c)$, where d is He d-line (588 nm); f, H F-line (486 nm) and c, H C-line (656 nm). (These are standard optical wavelengths.)

References: Glass Research Institute and British Ceramics R.A., manufacturers' leaflets.

M	Molar mass.
ρ	Density at 298 K.
α	Cubic expansivity (at 293 K).
T_m	Melting temperature (at 1 atm).
T_b	Boiling temperature (at 1 atm).

p_{sat}	Saturation vapour pressure (at 298 K).
$\Delta h_m^\ominus$	Specific standard enthalpy change (latent heat) of fusion (at T_m and 1 atm). $u(h_m^\ominus) = 10^4\,\mathrm{J\,kg^{-1}}$.
$\Delta h_b^\ominus$	Specific standard enthalpy change (latent heat) of vaporization (at T_b and 1 atm). $u(h_m^\ominus) = 10^4\,\mathrm{J\,kg^{-1}}$.

	Liquid	M $\mathrm{g\,mol^{-1}}$	ρ $\mathrm{g\,cm^{-3}}$	$\alpha^\dagger$ $10^{-3}\,\mathrm{K^{-1}}$	T_m K	T_b K	p_{sat} Pa	$\Delta h_m^\ominus$ $u(h_m^\ominus)$	$\Delta h_b^\ominus$ $u(h_m^\ominus)$
1	Water	18.0	1.00	0.21	273.1	373.1	2261	33.44	226.1
2	Heavy water	20.0	1.10		277.0	374.6		31.70	
3	Mercury	200.6	13.59	0.18	234.3	630.1		1.15	29.5
4	Pentane	72.2	0.63	1.61	143.1	309.2	56392		35.9
5	Hexane	86.2	0.66		178.1	342.1	16093		33.2
6	Heptane	100.2	0.68		182.5	371.5	4655		32.5
7	Octane	114.2	0.70		216.3	398.8	1330		29.2
8	Dichloromethane	84.9	1.32	1.37	178.0	313.1	44954	5.42	33.0
9	Trichloromethane	119.4	1.48	1.27	209.6	334.8	20615	7.79	24.9
10	Tetrachloromethane	153.8	1.59	1.24	250.1	349.6	11571	1.69	19.5
11	Dibromomethane	173.9	2.50		220.6	370.1	4522		
12	Carbon disulphide	76.1	1.26	1.22	162.2	319.5	39235	5.77	35.2
13	Methanol	32.0	0.79	1.20	179.2	338.1	12502	6.91	110.3
14	Ethanol	46.1	0.79	1.12	155.8	351.6	5586	10.89	83.9
15	Propan-1-ol	60.1	0.80	0.96	146.6	370.5	1862		68.7
16	Propane-1,2,3-triol (glycerol)	92.1	1.26	0.51	293.1	563.1	133	19.87	
17	Benzene	78.1	0.88	1.24	278.6	353.2	9975	12.70	39.4
18	Cyclohexane	84.2	0.78		279.6	353.8	10241	3.18	35.7
19	Phenylamine (aniline)	93.1	1.02	0.86	266.8	457.1	133	8.81	43.4
20	Methylbenzene (toluene)	92.1	0.87	1.07	178.1	383.7	2926		35.9
21	1,2-dimethylbenzene[A]	106.2	0.88	0.97	247.9	417.5	665	12.81	34.7
22	1,3-dimethylbenzene[A]	106.2	0.86	1.01	225.2	412.2	798	10.89	34.3
23	1,4-dimethylbenzene[A]	106.2	0.86	1.01	286.4	411.4	931	16.48	33.9
24	Ethanoic (acetic) acid	60.1	1.05	1.07	289.7	391.0	1596	19.47	39.4
25	Propanone (acetone)	58.1	0.79	1.49	177.8	329.3	23541	9.79	52.2
26	Ethoxyethane (ether)	74.1	0.71	1.66	156.9	307.6	57855		37.2
27	Turpentine	136.2	$0.86^{273\,K}$	0.97	263.2	429.2			28.7
28	Silicone oil[B]	163.3	$0.76^{273\,K}$	1.60	205.2	372.7			

$\dagger$ Many discrepancies between sources.
[A] o, m, and p-xylene respectively.
[B] Dimethyl silicone, low viscosity oil.

References: American Institute of Physics Handbook, Dreisbach, **Kaye**, Thermodynamics Research Centre, Weast.

κ Isothermal compressibility (293 K). $u(\kappa) = 10^{11}$ Pa^{-1}.
c_p Specific heat capacity. $u(c_p) = $ J g^{-1} K^{-1}.
λ Thermal conductivity. $u(\lambda) = 0.1$ W m^{-1} K^{-1}.
c Speed of sound (293 K).
η Viscosity (298 K). $u(\eta) = 10^{-4}$ N s m^{-2}.

γ Surface tension (293 K). $u(\gamma) = 10^{-2}$ N m^{-1}.
ε Relative permittivity (dielectric constant) (293 K, 0 Hz).
n Refractive index.

	Formula	$\kappa^{\dagger}$ u(κ)	c_p u(c_p)	$\lambda^{\dagger}$ u(λ)	c m s^{-1}	$\eta^{\dagger A}$ u(η)	γ^{B} u(γ)	ε	n(589 nm)
1	H$_2$O	0.46	4.17	6.0	1483	8.91	7.28	80.10	1.3325
2	D$_2$O	0.47	4.10	5.8	1384			79.80	1.3280
3	Hg		0.14	80.3	1451	15.50	40.70	—	—
4	CH$_3$(CH$_2$)$_3$CH$_3$	0.32		1.4	1044	2.24	1.60	1.84	1.3575
5	CH$_3$(CH$_2$)$_4$CH$_3$	1.54	2.26	1.3	1085	2.98	1.84	1.89	1.3749
6	CH$_3$(CH$_2$)$_5$CH$_3$	1.44	2.05	1.4	1161	3.96	2.03	1.92	1.3876
7	CH$_3$(CH$_2$)$_6$CH$_3$	1.16	2.22	1.5	1192	6.14	2.48	1.95	1.3974
8	CH$_2$Cl$_2$	0.97	1.22		1064	4.25	2.80	9.08	1.4211
9	CHCl$_3$	1.01	0.98	1.2	995	5.42	2.71^S	4.81	1.4429
10	CCl$_4$	1.05	0.84	1.1	938	8.80	2.69	2.24	1.4601
11	CH$_2$Br$_2$	0.65			971			7.73	1.5389
12	CS$_2$	0.93	0.99	1.6	1166	3.63	3.23	2.64	
13	CH$_3$OH	1.23	2.53	2.0	1122	5.53	2.26	33.00	1.3280
14	CH$_3$CH$_2$OH	1.11	2.41	1.7	1177	10.60	2.23	25.70	1.3610
15	CH$_3$CH$_2$CH$_2$OH	1.00	2.41	1.6	1223	22.70	2.30	20.10	1.3860
16	CH$_2$OHCHOHCH$_2$OH	0.21	2.42	2.9	1930	9420	6.30	42.50	1.4746
17	C$_6$H$_6$	0.96	1.70	1.5	1321	6.01	2.888^S	2.28^S	1.5010
18	C$_6$H$_{12}$	1.10	1.83		1278	8.95	2.50	2.02^S	1.4260
19	C$_6$H$_5$NH$_2$	0.56	2.05	1.7	1659	3.71	4.29	6.89	
20	C$_6$H$_5$CH$_3$	0.89	1.68	1.8	1322	5.50	2.84	2.39	1.4970
21	C$_6$H$_4$(CH$_3$)$_2$		1.73	1.4	1352	7.54	3.01	2.57	1.5060
22	C$_6$H$_4$(CH$_3$)$_2$	0.85	1.68	1.6		5.79	2.89	2.38	1.4970
23	C$_6$H$_4$(CH$_3$)$_2$		1.67			6.03	2.84	2.27	1.4960
24	CH$_3$CO$_2$H	0.91	2.03	1.7	1585	11.55	2.76	6.15	1.3719
25	CH$_3$COCH$_3$	1.27	2.17	1.9	1197	3.16	2.37	21.30	1.3587
26	C$_2$H$_5$OC$_2$H$_5$	1.87	2.28	1.4		2.22	1.696^S	4.34	1.3524
27	C$_{10}$H$_{16}$	1.28	1.75		1225		2.70		1.4700
28	CH$_3$Si(CH$_3$)$_2$OSi(CH$_3$)$_3$		1.37	1.0	795	4.95	1.60	2.18	1.3750

† Many discrepancies between sources.
^{A} Decreases rapidly with temperature, increases with pressure (doubles at about 10^8 Pa).
^{B} γ decreases rapidly with temperature. For many liquids, the empirical Eotvos relation holds:
$\mathrm{d}\{\gamma(M/\rho)^{2/3}\}/\mathrm{d}T = -2.12 \times 10^{-7}$ J mol$^{-2/3}$ K^{-1}, which has been used to estimate M.
^{S} Standard value for calibrating instruments.

7

M Molar mass.
T_b Boiling temperature (at 1 atm).
c_p Specific heat capacity (at 1 atm).
γ Heat capacity ratio c_p/c_V.
η Viscosity. $u(\eta) = 10^{-5}\,\text{N s m}^{-2}$.
α_V Volume expansivity.
α_p Temperature coefficient of pressure.
z Compressibility factor pV_m/RT, where V_m is the molar volume.
T_c Critical temperature.
p_c Critical pressure.
ρ_c Critical density. $u(\rho_c) = 10^2\,\text{kg m}^{-3}$.
l Mean free path.
d Molecular diameter (derived from viscosity measurements).

Gas	M / g mol^{-1}	T_b / K	c_p(273 K) / 0.1 J g^{-1} K^{-1}	γ(273 K)	η(273 K) $u(\eta)$	α_V(273 K) / 10^{-3} K^{-1}	α_p(273 K) / 10^{-3} K^{-1}	z(273 K, 1 atm)	T_c / K	p_c / 10^5 Pa	ρ_c $u(\rho_c)$	l(1 atm) / 10^{-8} m	d / 10^{-10} m
Ideal monatomic gas				1.67	0			1 (exactly)		—	—	∞	0
Air			[A]	1.40	1.71	3.66[B]	3.66[B]	0.99956	132	37.7	3.11	5.98	3.74
Oxygen (O_2)	32	90	9.09	1.40	1.92	3.67	3.67	0.99922	155	50.6	4.10	6.33	3.54
Nitrogen (N_2)	28	77	10.36	1.40	1.66	4.86	3.67	0.99968	126	33.9	3.11	5.88	3.75
Hydrogen (H_2)	2	20	141.50	1.41	0.84	3.67	3.66	1.0006	33	12.9	0.31	11.1	2.97
Helium (He)	4	4	52.25[C]	1.63[C]	1.86	3.66	3.66		5	2.3	0.69	17.4	2.58
Neon (Ne)	20	27	10.30	1.64	2.97	3.66	3.66		44	27.2	4.84	12.4	2.79
Argon (Ar)	40	87	5.24	1.67	2.10	3.68	3.67	0.99921	151	48.5	5.31	6.26	3.42
Chlorine (Cl_2)	71	238	4.81	1.36	1.23	3.83	3.80		417	76.9	5.73	2.74	4.40
Carbon monoxide (CO)	28	82	10.36	1.40	1.66	3.67	3.67		133	34.8	3.01	5.86	3.71
Carbon dioxide (CO_2)	44	195	8.32	1.30	1.38	3.74	3.73	0.99479	304	73.6	4.68	3.90	3.90
Sulphur dioxide (SO_2)	64	263	6.33	1.26[†]	1.17	3.90	3.84		431	78.6	5.24	2.74	4.29
Methane (CH_4)	16	109	22.06	1.31	1.03	3.68	3.68	0.9984[D]	191	46.2	1.62	4.81	3.80
Ethane (C_2H_6)	30	185	16.15	1.22	0.85			0.9926[D]	305	48.8	2.03		4.42
Propane (C_3H_8)	44	231	$2.23^{-43\,°C}$	1.13	0.80			0.9848[D]	370	42.4	2.20		5.08
Butane (C_4H_{10})	58	273	$2.27^{-5\,°C}$	$1.11^{15\,°C}$	$0.83^{16\,°C}$			0.9712[D]	425	37.8	2.28		5.00
2-methylpropane (C_4H_{10})	58	261	$2.20^{-16\,°C}$	1.34[†]	$0.76^{23\,°C}$			0.9731[D]	408	36.4	2.21		5.34
Ammonia (NH_3)	17	240	21.88	1.32	0.92	3.77	3.79		405	112	2.35	5.83	2.97
Hydrogen sulphide (H_2S)	34	212	10.00	1.32	1.17	3.77	3.76		374	89.8	3.49		
Ethene (ethylene) (C_2H_4)	28	169	15.02	1.26	0.91	3.72	3.74		283	51.0	2.27	3.43	4.23
Ethyne (acetylene) (C_2H_2)	26	189	$16.04^{15\,°C}$	1.26	0.94	3.74	3.73		309	62.2[†]	2.31		4.22
Dinitrogen oxide (N_2O)	44	185	8.25	1.30	1.35	3.73[†]	3.72[†]		310	72.4	4.59[†]	3.87	3.88
Nitrogen oxide (NO)	30	121	9.70	1.39	1.78	3.67	3.67		180	65.6	5.20		3.47
Nitrogen dioxide (NO_2)	46	294	6.80	1.31					431	101	5.60		
Refrigerant 12 (Freon 12) (CCl_2F_2)	121	243	$0.61^{30\,°C}$	1.14	$1.27^{30\,°C}$				385	41.0	5.55		
Refrigerant 11 (Freon 11) (CCl_3F)	137	297	0.67	1.14	$1.14^{30\,°C}$				471	43.6	5.54		

Note. Specific standard enthalpies of vaporization, $\Delta h_b^{\ominus}(T_b, 1\,\text{atm})$/J g^{-1}: H_2 450, O_2 213, He 21, CO_2 607, NH_3 1376, CCl_2F_2 165, CCl_3F 182.
† Discrepancy between sources.   [A] 207.8 g mol^{-1}/M.   [B] 1000/273.
[C] −180 °C.   [D] 300 K.

References: as Table 7.9.

z Compressibility factor $= pV_m/RT$.

B_V, C_V, D_V Virial coefficients in $pV_m/RT = 1 + B_V(T)/V_m + C_V(T)/V_m^2 + D_V(T)/V_m^3$. U cm³ mol⁻¹.

HYDROGEN

T/K	1 atm	4 atm	7 atm	10 atm	40 atm	70 atm	100 atm	B_V/U	C_V/U^2	$D_V/10^3 U^3$
				z						
40	0.9845	0.9362	0.8853	0.8317				−51.52	1400	−10.4
100	0.9998	0.9992	0.9987	0.9983	1.0029	1.0222	1.0560	−1.90	412	13.0
200	1.0007	1.0028	1.0048	1.0068	1.0283	1.0513	1.0760	11.93	254	8.85
300	1.0006	1.0024	1.0042	1.0059	1.0238	1.0420	1.0607	15.01	250	6.00
400	1.0005	1.0020	1.0034	1.0048	1.0193	1.0339	1.0486			
500	1.0004	1.0016	1.0028	1.0040	1.0160	1.0280	1.0400			
600	1.0003	1.0012	1.0023	1.0034	1.0136	1.0237	1.0337			

Critical constants: $T_c = 32.99$ K, $p_c = 1.294$ MPa, $\rho_c = 0.031$ g cm⁻³, $z_c = 0.30$.
Van der Waals constants: $a = 0.0247$ Pa m⁶ mol⁻², $b = 26.7 \times 10^6$ m³ mol⁻¹.

DRY AIR

T/K	1 atm	4 atm	7 atm	10 atm	40 atm	70 atm	100 atm	B_V/U	C_V/U^2	D_V/U^3
100	0.9809							−153.15	−3253.5	9.40
200	0.9977	0.9907	0.9837	0.9767	0.9080	0.8481	0.8105	−38.24	1323.5	5.46
300	0.9997	0.9988	0.9980	0.9972	0.9914	0.9900	0.9933	−7.480	1288.5	3.46
400	1.0002	1.0008	1.0014	1.0021	1.0095	1.0188	1.0299	6.367	1194.2	2.16
500	1.0003	1.0014	1.0024	1.0035	1.0145	1.0265	1.0393	14.048	1119.2	1.40
1000	1.0003	1.0013	1.0023	1.0033	1.0133	1.0233	1.0333	27.129	904.3	
2000	1.0004	1.0009	1.0014	1.0020	1.0076	1.0132	1.0188			
3000	1.0252	1.0133	1.0107	1.0095	1.0092	1.0119	1.0151			

Critical constants: $T_c = 132.45$ K, $p_c = 3.77$ MPa, $\rho_c = 0.311$ g cm⁻³, $z_c = 0.29$.
Van der Waals constants: $a = 0.14$ Pa m⁶ mol⁻², $b = 39.1 \times 10^6$ m³ mol⁻¹ (for N_2).

CARBON DIOXIDE

T/K	1 atm	4 atm	7 atm	10 atm	40 atm	70 atm	100 atm
300	0.9950	0.9798	0.9644	0.9486	0.7611		
400	0.9982	0.9927	0.9871	0.9815	0.9252	0.8697	0.8155
500	0.9993	0.9971	0.9950	0.9928	0.9721	0.9531	0.9365
600	0.9998	0.9990	0.9983	0.9976	0.9916	0.9874	0.9850
700	1.0000	0.9999	0.9999	1.0000	1.0008	1.0031	1.0068
800	1.0001	1.0004	1.0008	1.0011	1.0054	1.0108	1.0172
900	1.0001	1.0007	1.0012	1.0018	1.0079	1.0147	1.0224
1000	1.0002	1.0008	1.0015	1.0022	1.0092	1.0167	1.0248
1500	1.0002	1.0010	1.0017	1.0025	1.0100	1.0176	1.0253

Critical constants:
$T_c = 304.2$ K
$p_c = 7.36$ MPa
$\rho_c = 0.468$ g cm⁻³
$z_c = 0.274$
Van der Waals constants:
$a = 0.3636$ Pa m⁶ mol⁻²
$b = 42.67 \times 10^6$ m³ mol⁻¹

BUTANE

T/K	1 atm	4 atm	7 atm	10 atm	40 atm	70 atm	100 atm	B_V/U
300	0.9712	0.8606ᴴ						
350	0.9840	0.9274	0.8346	0.7119ᴴ				
400	0.9888	0.9553	0.9063	0.8541				
450	0.9920	0.9680	0.9350	0.9908	0.5949	0.3358	0.4095	0.9816
500	0.9944	0.9773	0.9543	0.9313	0.7644	0.5693	0.5214	0.9745
600	0.9972	0.9887	0.9775	0.9664	0.8913	0.8193	0.7780	1.0051
700	0.9985	0.9940	0.9882	0.9825	0.9464	0.9164	0.9026	1.0674
800	0.9992	0.9968	0.9938	0.9909	0.9743	0.9643	0.9637	1.1199
900	0.9997	0.9985	0.9971	0.9958	0.9902	0.9909	0.9975	1.1545
1000	0.9999	0.9996	0.9992	0.9989	0.9998	1.0066	1.0173	1.1751
1500	1.0003	1.0013	1.0026	1.0040	1.0151	1.0304	1.0458	1.1888

Critical constants:
$T_c = 425.16$ K
$p_c = 3.784$ M Pa
$\rho_c = 0.228$ g cm⁻³
$z_c = 0.274$
Van der Waals constants:
$a = 1.466$ Pa m⁶ mol⁻²
$b = 122.6 \times 10^6$ m³ mol⁻¹
ᴴ Liquid.

References: as Table 7.9.

Reference pressure for boiling and freezing temperatures is 101.325 kPa (1 atm). Defining points for the 1968 International Practical Temperature Scale (IPTS-68) are named in bold type.

	T_{68}/K	t_{68}/°C
Absolute zero (unattainable)	0	−273.15
Helium boils	4.215	−268.935
Triple point of hydrogen[A]	**13.81**	**−259.34**
Hydrogen[A] **vapour pressure = 33 330.6 Pa**	**17.042**	**−256.108**
Hydrogen[A] **boils**	**20.28**	**−252.87**
Neon boils	27.102	−246.048
Triple point of oxygen	**54.361**	**−218.789**
Oxygen boils	**90.188**	**−182.962**
Ethanol melts	161	−112
Carbon dioxide sublimes[S]	194.67	−78.48
Lowest recorded land surface temperature	205	−68
($CaCl_2 + 6H_2O$): ice (1:0.7) freezing mixture	218.3	−54.9
Mercury melts[S]	234.29	−38.86
Ethylene glycol: water 50:50 (antifreeze) freezes	236.7	−36.5
Triple point of water (definition of K)	**273.16**	**0.01**
Room temperature	288 to 293	15 to 20
Standard for thermochemistry	298.15	25.0
Hot weather	303	30
Highest recorded land temperature	330	57
Ethanol boils	351.6	78.4
Water boils } alternative defining points	**373.15**	**100.00**
Tin freezes	**505.1181**	**231.9681**
Mercury boils[S]	629.81	356.66

	T_{68}/K	t_{68}/°C
Freezing point of zinc	**692.73**	**419.58**
Dull red (black body)	800	530
Bright red (black body)	1180	900
Freezing point of silver	**1235.08**	**961.93**
Freezing point of gold	**1337.58**	**1064.43**
White heat	1500 and above	1300 and above
Iron melts	1808	1535
Bunsen burner flame (town gas)	2033	1760
Oxy-hydrogen flame	3073	2800
Tungsten filament	3100 to 3300	2800 to 3000
Tungsten melts[S]	3660	3387
Electric arc	3700	3427
Oxy-acetylene flame	3773	3500
Carbon sublimes	5100	4827
Tungsten vaporizes	5570	5297
Surface of Sun[B]	6000	5700
Nitroglycerine explosion (calc.)	7000	6700
Hottest star (surface)	25000	
Solar corona	10^6	
Stellar interior, nuclear bomb	10^8	

[A] Hydrogen must achieve equilibrium between para and ortho molecules.   [B] Varies with method of measurement.   [S] Secondary standard.

References for Tables 7.12, 7.13, and 7.14: Kaye, Weast.

1 Below 5 K, helium vapor pressure. Scales of 1958 (^{4}He) and 1962 (^{3}He). p_{sat} saturated vapour pressure of helium.

p_{sat}/Pa	10^{-2}	10^{-1}	1.0	10	10^2	10^3	10^4	10^5	1.01×10^5
$T_{58}(^4$He)/K	0.549	0.643	0.771	0.953	1.23	1.67	2.48	4.20	4.125
$T_{62}(^3$He)/K	0.228	0.276	0.345	0.452	0.632	0.966	1.66	3.18	3.190

2 13.81 K to 903.89 K, platinum resistance thermometer. 4 different formulae used over different ranges.
3 903.89 K to 1337.58 K, Pt/Pt$_{90}$Rh$_{10}$ thermocouple. E.M.F. of thermocouple $E(T_{68}) = a + bT_{68} + cT_{68}^2$. a, b, and c are constants.
4 Above 1337.58 K, disappearing filament optical pyrometer.

TEMPERATURE SCALES AND TEMPERATURE MEASUREMENT | **7·14**

T, θ Temperature. R Gas constant. k Boltzmann constant.
$\rho(T)$ Electrical resistivity, for Pt, Cu, and W respectively, at temperature T.
U Thermocouple e.m.f. for Pt/Pt$_{90}$Rh$_{10}$, copper/constantan, and chromel/alumel respectively. Cold junction is at 273.15 K.

The figures given may be interpolated to obtain approximate thermometer calibrations.

$\dfrac{T}{K}$	$\dfrac{\theta_C}{°C}$	$\dfrac{\theta_F}{°F}$	$\dfrac{RT}{kJ\,mol^{-1}}$	$\dfrac{kT}{eV}$	$\dfrac{\rho_{Pt}(T)}{\rho_{Pt}(°C)}$	$\dfrac{\rho_{Cu}(T)}{\rho_{Cu}(°C)}$	$\dfrac{\rho_W(T)}{\rho_W(°C)}$	$\dfrac{U_{Pt/PtRh}}{mV}$	$\dfrac{U_{Cu/con}}{mV}$	$\dfrac{U_{ch/al}}{mV}$
73.2	−200	−328	0.61	0.006	0.177	0.117	0.122			
123.2	−150	−238	1.02	0.011					−4.60	−4.81
173.2	−100	−148	1.44	0.015	0.599	0.557			−3.35	−3.49
223.2	−50	−58	1.86	0.019					−1.81	−1.86
273.2	0	32	2.27	0.024	1.000	1.000	1.000			
323.2	50	122	2.69	0.028				0.30	2.03	2.02
373.2	100	212	3.10	0.032	1.392	1.431	1.490	0.64	4.28	4.10
423.2	150	302	3.52	0.036				1.03	6.70	6.13
473.2	200	392	3.93	0.041	1.773	1.852		1.44	9.29	8.13
523.2	250	482	4.35	0.045				1.87	12.01	10.16
573.2	300	572	4.77	0.049	2.142	2.299	2.531	2.32	14.86	12.21
623.2	350	662	5.18	0.054				2.78	17.82	14.29
673.2	400	752	5.60	0.058	2.499	2.747		3.25	20.87	16.40
723.2	450	842	6.01	0.062				3.73		18.51
773.2	500	932	6.43	0.067	2.844	3.210	3.673	4.22		20.65
823.2	550	1022	6.84	0.071				4.72		22.78
873.2	600	1112	7.26	0.075	3.178	3.695		5.22		24.91
923.2	650	1202	7.68	0.080				5.74		27.03
973.2	700	1292	8.09	0.084	3.499	4.207	4.898	6.26		29.17
1023.2	750	1382	8.51	0.088				6.79		31.23
1073.2	800	1472	8.92	0.093	3.809	4.750		7.33		33.30
1123.2	850	1562	9.34	0.097				7.88		35.34
1173.2	900	1652	9.75	0.101	4.108	5.332		8.43		37.36
1223.2	950	1742	10.17	0.105				9.00		39.35
1273.2	1000	1832	10.58	0.110	4.395	5.959	6.735	9.57		41.31
1323.2	1050	1922	11.00	0.114				10.15		43.25
1373.2	1100	2012	11.42	0.118	4.672			10.74		45.16
1423.2	1150	2102	11.83	0.123				11.34		47.04
1473.2	1200	2192	12.25	0.127	4.937		7.959	11.94		48.89
1523.2	1250	2282	12.66	0.131				12.54		50.69
1573.2	1300	2372	13.08	0.136	5.190			13.14		52.46
1623.2	1350	2462	13.49	0.140				13.74		54.20
1673.2	1400	2552	13.91	0.144	5.431			14.34		
1723.2	1450	2642	14.33	0.149				14.94		
1773.2	1500	2732	14.74	0.153	5.660			15.53		
1823.2	1550	2822	15.16	0.157				16.12		
1873.2	1600	2912	15.57	0.161				16.72		
1923.2	1650	3002	15.99	0.166				17.31		
1973.2	1700	3092	16.40	0.170				17.89		
2023.2	1750	3182								

7

COPPER WIRE

D_b Diameter of bare conductor. Preferred sizes in bold.

D_i Diameter of conductor plus grade I (350 V breakdown) enamel insulation ([II] denotes grade II insulation).

R Resistance per unit length.

ε Minimum elongation on fracture.

s.w.g.   Near equivalent standard wire gauge (now obsolete).

D_s Diameter of quoted s.w.g. (given to illustrate the basis of s.w.g.).

D_b mm	D_i mm	$R(293\,\mathrm{K})$[1] $\Omega\,\mathrm{m}^{-1}$	ε %	s.w.g.	D_s in	D_b mm	D_i mm	$R(293\,\mathrm{K})$[1,2] $10^{-2}\,\Omega\,\mathrm{m}^{-1}$	ε %	s.w.g.	D_s in
0.016	0.020	85.75	—	—	—	**0.315**	0.352	22.12	23	30[3]	0.0124
0.020	0.025	54.88	—	50	0.0010	**0.400**	0.442	13.72	24	27	0.0164
0.025	0.031	35.12	—	49	0.0012	**0.500**	0.548	8.781	25	25	0.020
0.032	0.040	21.44	—	—	—	**0.630**	0.684	5.531	27	23	0.024
0.040	0.050	13.72	—	48	0.0016	**0.800**	0.861	3.430	28	21	0.032
0.050	0.062	8.781	10	47	0.0020	**1.000**	1.068	2.195	30	19	0.040
0.063	0.078	5.531	12	46	0.0024	**1.250**	1.325	1.405	31	18	0.048
0.080	0.098	3.430	14	44	0.0032	**1.600**	1.683	0.858	32	16	0.064
0.100	0.129[II]	2.195	16	42	0.0040	**2.000**	2.092	0.549	33	14	0.080
0.125	0.149	1.405	17	38	0.0048	**2.500**	2.631[II]	0.351	33	12	0.140
0.160	0.187	0.8575	19	37	0.0068	**3.150**	3.294[II]	0.221	34	10	0.128
0.200	0.230	0.5488	21	36	0.0076	**4.00**	4.160[II]	0.137	35	8	0.160
0.250	0.284	0.3512	22	33	0.0100	**5.00**	5.177[II]	0.088	36	6	0.192

[1] Tolerance $\pm 10\%$ on fine wires to $\pm 3\%$ on thick wires.   [2] Note change of scale when compared to lefthand column.   [3] Exact equivalence of metric and s.w.g. sizes.

INTERNATIONAL COLOUR CODE FOR 3-CORE FLEX

brown: **live**   blue: **neutral**   yellow/green: **earth**

(Old British standard   red: live   black: neutral   green (sometimes brown): earth)

INTERNATIONAL CODE FOR RESISTORS

1st letter (shows position of decimal point):   R ohms   K kilohms   M megohms

2nd letter (shows tolerance):   F $\pm 1\%$ G $\pm 2\%$ J $\pm 5\%$ K $\pm 10\%$ M $\pm 20\%$

Thus: 1ROM denotes $1.0\,\Omega \pm 20\%$ 100K0K denotes $100\,\mathrm{k}\Omega \pm 10\%$ 6K8G denotes $6.8\,\mathrm{k}\Omega \pm 2\%$

       4R7J denotes $4.7\,\Omega \pm 5\%$ 4M7F $4.7\,\mathrm{M}\Omega \pm 1\%$

PREFERRED VALUES FOR RESISTORS

Bold type: available in $\pm 5\%$, $\pm 10\%$, and $\pm 20\%$ tolerance ranges

Normal type: available in $\pm 5\%$, and $\pm 10\%$ tolerance ranges only

Italic type: available in $\pm 5\%$ tolerance range only

Figures are repeated over each decade from $0.22\,\Omega$ to $22\,\mathrm{M}\Omega$. Values outside this range are not always available.

10   *11*   12   *13*   **15**   *16*   18   *20*   **22**   *24*   27   *30*   **33**   *36*   39   *43*   **47**   *51*   56   *62*   **68**   *75*   82   *91*   **100**

References: British Standards Institute BS 1852 and BS 4516, Institution of Electrical Engineers.

Recommended symbols given in Table 1.2 are used, with a minimum of explanation. The list is intended only to refresh the memory. It is necessary to consult a textbook to find out the meaning of the formulae and the assumptions made in deriving them.

Motion and forces

Linear motion	$a = \mathrm{d}v/\mathrm{d}t = \mathrm{d}^2s/\mathrm{d}t^2$	
	$v = u + at$	Constant a only.
	$v^2 = u^2 + 2as$	Constant a only.
Circular motion	$v = r\omega$	
	$a = v^2/r$	
Force = rate of change of momentum	$F = \mathrm{d}(mv)/\mathrm{d}t$	
= mass × acceleration	$F = ma$	
Work = component of force × displacement	$W = F\Delta s$	
Kinetic energy	$E_k = \frac{1}{2}mv^2$	
Gravitational force	$F = Gm_1m_2/r^2$	Gravitational constant G, masses m_1 and m_2,
field	$g = Gm/r^2$	distance r. Magnitudes only are given, not
potential energy	$E_p = Gm_1m_2/r$	directions or signs.
potential	$V_g = Gm/r$	
potential energy difference	$\Delta E_p = mgh$	Uniform field strength g, height h.
Kepler's laws	$R^2\omega = \text{constant}$	Average distance from central body R, angular
	$GmT^2 = 4\pi R^3$	velocity ω, period of revolution about central body T.
Hubble's law	$v = \alpha r$	Speed of recession of distant galaxy v, distance of galaxy r.
Rotational motion		
moments of inertia	$I = \sum mr^2 = Mk^2$	In general.
	$I = mr^2/2$	For disc about perpendicular axis through centre.
	$I = mr^2/4$	For disc about diameter.
	$I = ml^2/12$	For bar about perpendicular axis through centre.
	$I = 2mr^2/3$	For hollow sphere about diameter.
	$I = 2mr^2/5$	For solid sphere about diameter.
perpendicular axes theorem	$I_z = I_x + I_y$	For plane lamina.
parallel axes theorem	$I_a = I_c + Mh^2$	I_c relates to ∥ axis through centre of mass.
kinetic energy	$E_k = I\omega^2/2$	
torque	$T = Fr = \mathrm{d}(I\omega)/\mathrm{d}t$	$r \perp F$
Simple harmonic motion	$a = -\omega^2 x$	Acceleration $a = \mathrm{d}^2x/\mathrm{d}t^2$, displacement x, time t.
	$\omega^2 = k/m$	Force per unit displacement k, mass m.
period (frequency = $1/T$)	$T = 2\pi/\omega$	
	$\quad = 2\pi\sqrt{l/g}$	For simple pendulum.
	$\quad = 2\pi\sqrt{I/\tau}$	For torsional oscillations.
	$\quad = 2\pi\sqrt{m/k}$	For mass on spring.
	$\omega = 2\pi f$	Frequency f.
displacement	$x = A\cos\omega t$	Amplitude A.
Spring obeying Hooke's law	$F = kx$	Restoring force F, force constant k,
stored energy	$E = \frac{1}{2}kx^2 = \frac{1}{2}Fx$	displacement x.
Stress	$\sigma = F/A$	
Strain	$\varepsilon = \Delta x/x_0$	

8

Tensile (Young) modulus	$E = \sigma/\varepsilon$	
Stored energy density	$E_p = \sigma\varepsilon/2 = E(\Delta x)^2/2x_0$	
Poisson ratio	$\mu = -\dfrac{\varepsilon(\text{lateral})}{\varepsilon(\text{longitudinal})}$	
Shear stress	$\tau = F/A$	F acts parallel to 'surface'.
Shear strain	$\gamma = \dfrac{\Delta x(\|F)}{x(\perp F)} = \Delta\theta$	θ is angle of shear.
Shear modulus	$G = \tau/\gamma$	
Bulk modulus	$K = -p/(\Delta V/V_0)$	

Electricity

	$V = IR$	$\begin{cases} \text{Potential difference } V, \text{ current } I, \\ \text{resistance } R. \end{cases}$
Power	$P = IV = I^2R = V^2/R$	
Resistance	$R = \rho L/A$	Resistivity ρ, length L, area A.
series connection	$R = R_1 + R_2 + \cdots$	
parallel connection	$1/R = 1/R_1 + 1/R_2 + \cdots$	
Capacitance	$C = Q/V$	Capacitance C, charge Q.
	$C = \varepsilon A/d = \varepsilon_r\varepsilon_0 A/d$	Parallel plates, area A, spacing D, permittivity ε, permittivity of vacuum ε_0, relative permittivity ε_r.
parallel connection	$C = C_1 + C_2 + \cdots$	
series connection	$1/C = 1/C_1 + 1/C_2 + \cdots$	
Decay of charge	$Q = Q_0 e^{-t/RC}$	Initial charge Q_0, time t.
Energy stored by capacitor	$E = \frac{1}{2}QV$	
	$(= \frac{1}{2}CV^2 = \frac{1}{2}Q^2/C)$	
Electric field	$E = \text{force per unit charge}$	
	$\quad = -(\text{potential gradient})$	
	$E = -\Delta V/\Delta x$	Potential difference ΔV, distance Δx.
uniform field only	$E = V/d$	Potential difference V, distance d.
Point charges in vacuum	$E = Q/4\pi\varepsilon_0 r^2$	
force	$F = Q_1Q_2/4\pi\varepsilon_0 r^2$	
potential	$V = Q/4\pi\varepsilon_0 r$	
potential energy	$E_p = Q_1Q_2/4\pi\varepsilon_0 r$	
Conductance	$G = 1/R = \kappa A/L$	Conductivity κ, length L, area A.

Reactive circuits

Time constant	$\tau = RC$	Resistance R, capacitance C.
	$\quad = L/R$	Inductance L.
Reactance	$X = 1/2\pi fC = 1/\omega C$	Frequency f, angular frequency ω.
	$X = 2\pi fL = \omega L$	
Resonant frequency	$2\pi f = \omega = 1/\sqrt{LC}$	
Force on current	$F = BIL \sin\theta$	Flux density B, length L, angle θ.
Force on moving charge	$F = BQv \sin\theta$	Charge Q, velocity v.
Force between 2 parallel wires carrying electric current in vacuum	$F = I_1I_2\mu_0 l/2\pi d$	Currents I_1, I_2, wire lengths l, separation d, permeability of vacuum μ_0.
Induced e.m.f. = rate of change of flux linked	$E = -\mathrm{d}\Phi/\mathrm{d}t$	Flux Φ.
Round a closed loop	$\oint B\,\mathrm{d}s = \mu_0 I$	Length along field s.
Flux × reluctance = current turns	$\Phi R = n$	Reluctance R, current turns n.

Flux	$\Phi = LI$	Flux Φ, self inductance L.
	$V = L\,dI/dt$	Voltage V needed to maintain rate of change of current dI/dt.

Flux density in vacuum
from currents

long solenoid	$B = \mu_0 NI/L$	Turns N, length L.
long straight wire	$B = \mu_0 I/2\pi r$	Radial distance r.
circular coil	$B = \mu_0 NI/2r$	Radius r, B at centre of coil.
Torque on dipole	$T = mB \sin\theta$	Magnetic moment m.
Torque on coil	$T = ANIB \sin\theta$	Angle between normal to coil (RH screw + ve) and field θ.
Mutual inductance	$\Phi_2 = M_{12}I_1 = M_{21}I_1$	Flux linking circuit 2 Φ_2.
Stored energy	$E_p = LI^2/2$	

Varying currents and a.c.

RC circuit	$Q = CV_0(1 - e^{-t/RC})$	Growth of charge, constant V_0.
	$Q = CV_0 e^{-t/RC} = e^{-t/\tau}$	Decay of charge, initial V_0.
RL circuit	$I = (V_0/R)(1 - e^{-Rt/L})$	Growth of current, constant V_0.
	$I = I_0 e^{-Rt/L} = I_0 e^{-t/\tau}$	Decay of current, initial I_0.
LRC circuit	$V = L(dI/dt) + RI + Q/C$	
a.c. series circuit	$E = E_0 \cos\omega t$	
	$I = (E_0/Z)\cos(\omega t - \phi)$	Impedance $Z = (R^2 + (\omega L - 1/\omega C)^2)^{1/2}$, phase angle ϕ, $\tan\phi = (\omega L - 1/\omega C)/R$.
	$I_{\text{r.m.s.}} = E_{\text{r.m.s.}}/Z$	E and I are often used to denote r.m.s. values without subscript.
	$E_{\text{r.m.s.}} = E_0/\sqrt{2}$	Peak value e.m.f. E_0.
	$= 0.707 E_0$	
Resonant frequency	$f_0 = 1/2\pi\sqrt{LC}$	
	$Z(v_0) = R$	

Waves and light

	$c = f\lambda$	Wave speed c, frequency f, wavelength λ.
Wave speeds: electromagnetic	$c = 1/\sqrt{\varepsilon_0 \mu_0}$	
sound, solid rod	$c = \sqrt{E/\rho}$	Young modulus E, density ρ.
sound, gas	$c = \sqrt{\gamma p/\rho}$	Pressure p, constant γ.
row of equal masses and springs	$c = x\sqrt{k/m}$	Spacing x, force per unit displacement k, mass m.
transverse wave on string	$c = \sqrt{F/\mu}$	Tension F, mass per unit length μ.
Diffraction: narrow slit	$\sin\theta = \lambda/d$	Angle of first minimum θ, width d.
circular hole	$\sin\theta = 1.22\lambda/d$	Diameter d.
Diffraction grating	$n\lambda = s\sin\theta$	Order n, slit spacing s, angles of maxima θ.
	$\Delta\lambda/\lambda = 1/nN$	Wavelength difference resolved $\Delta\lambda$, order n, number of slits N.
Young's fringes	$\Delta x/\lambda = D/d$	Separation of fringes Δx, distance from slits to screen D, distance between slits d.
Bragg diffraction (first order)	$\lambda = 2d\sin\theta$	Layer spacing d, angle of strong 'reflection' θ.
Doppler effect	$f_1/f_2 = c/(c - v_s)$	Moving source $\quad\left.\right\{ v_s$ and v_o in same direction
	$= (c - v_0)/c$	Moving observer $\left.\right\}$ as c.

8

Radiation

Wien's law for black body	$\lambda_{\max} T = 2.9 \times 10^{-3}\,\text{m K}$	c constant
Kirchoff's law	$\varepsilon(\lambda)/a(\lambda) = dQ/d\lambda = c$	$\begin{cases} c_1 = 2\pi hc^2 = 3.74 \times 10^{-16}\,\text{W m}^2 \\ c_2 = hc/k = 1.44 \times 10^{-2}\,\text{m K}. \end{cases}$
Planck's law	$E_\lambda\, d\lambda = c_1 \lambda^{-5}\, d\lambda (e^{c_2/\lambda T} - 1)^{-1}$	

Atomic physics

Radioactivity	$dN/dt = -\lambda N$	Number N, time t, decay constant λ.
	$N = N_0 e^{-\lambda t}$	Initial number N_0.
Half-life	$T_{1/2} = (\ln 2)/\lambda = 0.693/\lambda$	
Photon energy	$E = hf$	Planck constant h, frequency f.
Maximum kinetic energy of photo-electron	$E_{\max} = hf - \phi$	Work function ϕ.
De Broglie wavelength	$\lambda = h/p = h/mv$	Momentum $p = mv$.
Mass-energy relation	$E = mc^2$	Speed of light c, mass associated with this energy m.
Heisenberg uncertainty relation	$\Delta x\, \Delta p \geqslant h/2\pi$	Δx displacement uncertainty, Δp momentum uncertainty.

Gas laws

Ideal gas (molecular formula X)	$pV = nRT$	Gas constant[A] R, amount of X n.
	$pV = nLkT$	Avogadro constant L, Boltzmann constant[A] k.
	$pV = \frac{1}{3}Nm\overline{c^2}$	Number of molecules N, mass of molecule m, mean square speed $\overline{c^2}$.
Van der Waals equation	$(p + an^2/V^2)(V/n - b) = RT$	a, b are constants.
Dalton's law of partial pressure	$p = p_A + p_B + \cdots$	Pressures of gases existing on own p_A, $p_B \cdots$
Raoult's law	$p = p^{\circ}N/(n + N)$	Vapour pressure of solution p, of pure solvent p°; amount of solvent N, of involatile solute n.
Osmotic pressure	$\Pi V = nRT$ or $\Pi = RT[B]$	Concentration of solute B [B].

Entropy, energy, enthalpy

Entropy	$S = k \ln W$	Boltzmann constant k, number of molecular arrangements W.
	$\Delta S = Q_{\text{reversible}}/T$	For any process involving a single isothermal.
System at equilibrium	$\Delta S_{\text{total}} = 0$	
	$\Delta S_{\text{system}} + \Delta S_{\text{surroundings}} = 0$	
	$\Delta S_{\text{surroundings}} = -\Delta H_{\text{system}}/T$	
Trouton's rule	$-\Delta H_b/T_b \approx -82\,\text{J mol}^{-1}\,\text{K}^{-1}$	For normal non-associated liquids.
Ideal gases	$\Delta S = -Lk \ln(p_2/p_1)$	Change of pressure from p_1 to p_2.
When a system changes at constant temperature	$\Delta S_{\text{total}} = \Delta S_{\text{surroundings}} + \Delta S_{\text{system}}$	
	$\Delta G = \Delta H - T\Delta S$	

[A] If p is expressed in atm and V in dm^3, unit handling is made easier by using $R = 0.082\,\text{atm dm}^3\,\text{K}^{-1}\,\text{mol}^{-1}$, and $k = 1.38 \times 10^{-23}\,\text{J K}^{-1}$; $1\,\text{atm dm}^3 = 1.013 \times 10^2\,\text{J}$.

Gibbs free energy changes	$\Delta G = -zFE_{cell}$	e.m.f. of cell E_{cell}.
	$\Delta G^{\ominus} = -RT\ln K_{c/c^{\ominus}}$	Standard Gibbs free energy change at concentration $c^{\ominus}$, $\Delta G^{\ominus}$.
ideal systems	$\Delta G^{\ominus} = -RT\ln K_{p/p^{\ominus}}$	Standard Gibbs free energy change at pressure $p^{\ominus}$, $\Delta G^{\ominus}$.
Boltzmann equation (factor)	$\ln x = -E/kT$	Ratio of number of molecules in one state to number in another state (lower in energy by E) x, Boltzmann constant k (1.38×10^{-23} J K^{-1}).
Average energy of molecule	$E_{average}$ is of the order kT	Detailed dependence of $E_{average}$ upon T depends on molecules concerned.
Specific heat capacity	$c = Q/m\,\Delta T$	Heat exchanged Q, mass m, temperature change ΔT.

Physical chemistry

Equilibrium law	$K_c = \dfrac{[C]^p_{eqm}[D]^q_{eqm}}{[A]^m_{eqm}[B]^n_{eqm}}$	For $mA + nB \rightleftharpoons pC + qD$.
	$K_{c/c^{\ominus}} = K_c/(c^{\ominus})^{p+q-m-n}$	
Gas phase equilibrium	$K_p = K_c(RT)^{p+q-m-n}$	
	$K_{p/p^{\ominus}} = K_p/(p^{\ominus})^{p+q-m-n}$	
pH of a solution	$pH = -\lg([H^+]/\text{mol dm}^{-3})$	
pH of a buffer solution	$pH = pK_a + \lg([\text{base}]_{eqm}/[\text{acid}]_{eqm})$	
Nernst equation	$E = E^{\ominus} + \dfrac{RT}{zF}\ln\dfrac{[\text{oxidized form}]}{[\text{reduced form}]}$	Electrode and standard electrode potential E and $E^{\ominus}$, number of electrons transferred in the reaction involving H_2 and H^+ z, gas constant R (8.314 J K^{-1} mol^{-1}), Faraday constant F (9.648×10^4 C mol^{-1}).
Reaction rates	$-d[A]/dt = k_0$ or $k_0t = [A]_0 - [A]_t$	Zero order.
	$-d[A]/dt = k_1[A]$ or $k_1t = \ln([A]_0/[A])$ and $t_{1/2} = \ln 2/k_1 = 0.69/k_1$	First order.
	$-d[A]/dt = k_2[A]^2$ or $k_2t = 1/[A] - 1/[A]_0$ and $t_{1/2} = 1/[A]_0k_2$	Second order.
Arrhenius equation	$k = Ae^{-E_a/RT}$	Activation energy E_a, pre-exponential factor A, gas constant R (8.314 J K^{-1} mol^{-1}), any unit of k u.
	$\ln(k/u) = \ln(A/u) - E_a/RT$	
Molar conductivity	$\Lambda = \kappa/c$	Electrolytic conductivity κ, concentration of electrolyte c.
Degree of ionization	$\alpha = \Lambda/\Lambda_0$	Limiting molar conductivity Λ_0.
Ostwald dilution law	$K = \alpha^2 c/(1-\alpha)$	Acid/base ionization constant K, concentration c.
Kohlrausch law (C_mA_n)	$\Lambda_0 = m\lambda_0(C) + n\lambda_0(A)$	Limiting molar ionic conductivities λ_0.

8

MATHEMATICAL FORMULAE

Differentials and integrals

Functions in column A are the differentials of those in column B. Functions in column B are the indefinite integrals of those in column A; integration constants are omitted. $a = $ constant.

A differential	B integral
anx^{n-1}	ax^n
ax^n	$ax^{n+1}/(n+1)$
x^{-1}	$\ln ax$
$a\,e^{ax}$	e^{ax}
$\cos x$	$\sin x$
$a\cos(ax)$	$\sin(ax)$
$\sin x$	$-\cos x$
$a\sin(ax)$	$-\cos(ax)$
$\sec^2 x$	$\tan x$
$a\sec^2(ax)$	$\tan(ax)$
$(1-x^2)^{-1/2}$	$\sin^{-1}x$ or $-\cos^{-1}x$
$(1+x^2)^{-1}$	$\tan^{-1}x$
$u(dv/dx)+v(du/dx)$	uv
$\{v(du/dx)-u(dv/dx)\}/v^2$	u/v
$\sin^2 x$	$x/2-(\sin 2x)/4$
$\cos^2 x$	$x/2+(\sin 2x)/4$

Trigonometric functions

$\sin\theta = y/r = 1/\mathrm{cosec}\,\theta \qquad \cos\theta = x/r = 1/\sec\theta \qquad \tan\theta = y/x = \sin\theta/\cos\theta = 1/\cot\theta$

$\cos^2\theta + \sin^2\theta = 1 \qquad\qquad 1+\tan^2\theta = \sec^2\theta \qquad\qquad 1+\cot^2\theta = \mathrm{cosec}^2\theta$

$\sin(\theta\pm\phi) = \sin\theta\cos\phi \pm \cos\theta\sin\phi \qquad \sin 2\theta = 2\sin\theta\cos\theta$

$\cos(\theta\pm\phi) = \cos\theta\cos\phi \mp \sin\theta\sin\phi \qquad \cos 2\theta = \cos^2\theta - \sin^2\theta = 2\cos^2\theta - 1 = 1 - 2\sin^2\theta$

$\tan(\theta\pm\phi) = (\tan\theta\pm\tan\phi)/(1\mp\tan\theta\tan\phi) \qquad \tan 2\theta = 2\tan\theta/(1-\tan^2\theta)$

$$\sin\theta \pm \sin\phi = 2\,\frac{\sin}{\cos}\left(\frac{\theta+\phi}{2}\right)\frac{\cos}{\sin}\left(\frac{\theta-\phi}{2}\right) \qquad \cos\theta \pm \cos\phi = \pm 2\,\frac{\cos}{\sin}\left(\frac{\theta+\phi}{2}\right)\frac{\cos}{\sin}\left(\frac{\theta-\phi}{2}\right)$$

Cosine rule $\quad a^2 = b^2 + c^2 - 2bc\cos A \qquad$ for any triangle

Sine rule $\qquad a/\sin A = b/\sin B = c/\sin C \qquad$ for any triangle

Series

$$e^x = 1 + x + x^2/2! + x^3/3! + \cdots$$
$$\sin x = x - x^3/3! + x^5/5! - \cdots$$
$$\cos x = 1 - x^2/2! + x^4/4! - \cdots$$

$$\ln(1 + x) = x - x^2/2 + x^3/3 - \cdots \qquad -1 < x \leqslant 1$$
$$(1 + x)^n = 1 + nx + n(n-1)x^2/2! \qquad |x| < 1. \text{ Series terminates for any } x \text{ if } n \text{ is a positive integer}$$
$$+ n(n-1)(n-2)x^3/3! + \cdots$$

Coordinate geometry

Straight line $\quad y = mx + c$
$$y - y_0 = m(x - x_0) \qquad \text{through } (x_0, y_0)$$
Circle $\qquad (x - a)^2 + (y - b)^2 = R^2 \quad$ centre at a, b

Algebra

$$ax^2 + bx + c = 0 \Rightarrow x = \{-b \pm \sqrt{b^2 - 4ac}\}/2a$$
$$a^2 - b^2 = (a + b)(a - b) \qquad (a \pm b)^2 = a^2 \pm 2ab + b^2 \qquad a^3 \mp b^3 = (a \mp b)(a^2 \pm ab + b^2)$$

Geometry

Perimeter of circle	$L = 2\pi r$	
Area of triangle	$S = \frac{1}{2}ah = \frac{1}{2}bc \sin A$	$a = $ base
Area of circle	$S = \pi r^2$	
Surface area of sphere	$S = 4\pi r^2$	
Surface area of cone	$S = \pi r(r + l)$	$l = $ slant height
Volume of sphere	$V = 4\pi r^3/3$	
Volume of cone	$V = \pi r^2 h/3$	

Centres of gravity

Circular arc	$CG = (r \sin \theta)/\theta$	⎧ along radius of symmetry.
Circular sector lamina	$CG = (2r \sin \theta)/3\theta$	⎩ θ is angle subtended at C
Triangular lamina (ABC)	$AG = 2AM/3$	M is midpoint of BC
Semicircular lamina	$CG = 4r/3\pi$	along radius of symmetry
Hollow cone (without base)	$VG = 2h/3$	⎧ on axis. V is vertex
Solid cone	$VG = 3h/4$	⎩
Hemisphere	$CG = 3r/8$	along radius of symmetry

8

References

AMERICAN SOCIETY FOR METALS *Metals handbook*. Volume I, 8th edn. 1961.

AMERICAN SOCIETY FOR TESTING MATERIALS *Special technical publication 48–J*. 1960. (One of a number of yearly supplements to *STP 48— X-ray diffraction data cards for chemical analysis*. 1941–55.)

American Institute of Physics Handbook. 3rd edn. McGraw, 1972.

ASSOCIATION FOR SCIENCE EDUCATION *Chemical nomenclature, symbols, and terminology*. 1979. (This is the second edition; a third edition is due for publication at the end of 1984.)

ASSOCIATION FOR SCIENCE EDUCATION *SI units, signs, symbols, and abbreviations*. 1981.

AULT, A. *Techniques and experiments for organic chemistry*. 3rd edn. Allyn and Bacon, 1979.

AYLWARD, G.H. and FINDLAY, T.J.V. *SI chemical data*. 2nd edn. Wiley, 1976.

BAMFORD, C.H. and TIPPER, C.F.H. (eds) *Comprehensive chemical kinetics*. Elsevier. (Many volumes published from 1969 onwards.)

BERNARD, M. and BUSNOT, F. *Chimie générale et minérale*. Aide-Memoire Dunod. Bordas, 1978. (In two volumes.)

BONDI, A. 'Van der Waals volumes and radii.' *J. Phys. Chem*, volume 68(3), 1964, page 441.

BS 1852 *Marking codes for resistors and capacitors*. British Standards Institution, 1975.

BS 4516 *Enamelled copper conductors polyvinyl acetyl base with high mechanical properties. Part 1: round wire*. British Standards Institution, 1969, 1981.

BS 5555 *Specification for SI units and recommendations for the use of their multiples and of certain other units*. British Standards Institution, 1981.

BS 5775 *Specification for quantities, units, and symbols*. Part O, 1–13. British Standards Institution, 1979–82.

CODATA *Recommended consistent values of the fundamental physical quantities*. CODATA Bulletin no. 11, 1973. International Council of Scientific Unions: Committee on Data for Science and Technology. Frankfurt.

COPPER DEVELOPMENT ASSOCIATION *Copper data*, no. 12, 1964.

COTTRELL, T.L. *The strengths of chemical bonds*. 2nd edn. Butterworth, 1958.

DE BETHUNE, A.J. *ET AL. Standard aqueous electrode potentials and temperature coefficients at 25°C*. Hampel, Illinois, 1964.

DOYLE, M.P. and MUNGALL, W.S. *Experimental organic chemistry*. Wiley, 1980.

DREISBACH, R.R. *Advances in Chemistry series* Volumes 15, 22, and 29, *Physical properties of chemical compounds*. American Chemical Society, 1955, 1959, and 1961.

GRAY, C.H. (ed.) *Laboratory handbook of toxic agents*. 2nd edn. Royal Institute of Chemistry, 1966.

HAMPEL, C.A. (ed.) *Rare metals handbook*. 2nd edn. Reinhold, 1961.

HOTOP, H. and LINEBERGER, W.C. 'Binding energies in atomic negative ions.' *J. Phys. Chem. Ref. Data*, volume 4(3), 1975, page 539.

HULTGREN, R. *Selected values of thermodynamic properties of metals and alloys*. Wiley, 1963.

INSTITUTION OF ELECTRICAL ENGINEERS *Regulations for electrical installations*. 15th edn. 1981.

International Encyclopaedia of Chemical Science. Van Nostrand, 1964.

IUPAC *Manual of symbols and terminology for physicochemical quantities and units*. 2nd revision. Pergamon, 1979. (International Union of Pure and Applied Chemistry. Physical Chemistry Division, Commission on Symbols, Terminology, and Units.)

JENKINS, H.D.B. and PRATT, K.F. 'On "basic" radii of simple and complex ions and the repulsion energy of ionic crystals.' *Proc. Roy. Soc. London*. A356, 115. 1977.

JENKINS, H.D.B. in WEAST, R.C. (ed.) 'Table of lattice energies.' *CRC handbook of chemistry and physics*. CRC Press, 1982.

JOHNSON. D.A. *Some thermodynamic aspects of inorganic chemistry*. 2nd edn. Cambridge University Press, 1982.

KAYE, G.W.C. and LABY, T.H. *Tables of physical and chemical constants and some mathematical functions*. 14th edn. Longman, 1973.

KUHN, H.G. *Atomic spectra*. 2nd edn. Longman, 1969.

LANDOLT, H. and BÖRNSTEIN, R. *Eigenschaften der Materie in Ihren Aggregatzuständen*, 4-Teil Kalorische Zustandgrossen. Springer-Verlag, 1961.

LATIMER, W.M. *The oxidation states of the elements and their potentials in aqueous solutions*. 2nd edn. Constable, 1952.

LEDERER, C.M. and SHIRLEY, V. *Table of isotopes*. 7th revised edn. Wiley, 1979.

LINKE, W.F. and SEIDELL, A. *Solubilities*. 4th edn. Van Nostrand, 1958 and 1965 (2 volumes).

McGLASHAN, M.L. Monographs for teachers, no. 15. *Physicochemical quantities and units: the grammar and spelling of physical chemistry*. 2nd edn. Royal Institute of Chemistry, 1971.

MARTIN, A. and HARBISON, S.A. *Introduction to radiation protection*. 2nd edn. Chapman and Hall, 1979.

MASSACHUSETTS INSTITUTE OF TECHNOLOGY (SPECTROSCOPY LABORATORY) *Wavelength tables*. M.I.T./Wiley, 1939.

MOORE, C.E. *Ionization potentials and ionization limits derived from the analyses of optical spectra*. United States Department of Commerce, 1970. NSRDS–NBS 34. Washington.

PARSONS, R. *Handbook of electrochemical constants*. Butterworth, 1959.

PAULING, L. *The nature of the chemical bond and the structure of molecules and crystals*. 3rd edn. Cornell University Press, 1960.

PEDLEY, J.B. and RYLANCE, J. *Sussex–NPL computer-analysed thermochemical data: organic and organometallic compounds*. University of Sussex, 1979.

PIETERS, H.A.J. and CREYGHTON, J.W. *Safety in the chemical laboratory*. 2nd edn. Butterworth, 1957.

ROYAL SOCIETY OF CHEMISTRY (SYMBOLS COMMITTEE) *Quantities, units and symbols: a report*. 2nd edn. 1975.

SHANNON, R.D. and PREWITT, C.T. 'Effective ionic radii in oxides and fluorides.' *Acta Crystallogr.*, volume B25, 1969, page 925.

SHANNON, R.D. and PREWITT, C.T. 'Revised values of effective ionic radii.' *Acta Crystallogr.*, volume B26, 1970, page 1046.

SIEGBAHN, K. (ed.) α, β, *and γ-ray spectroscopy*. North Holland, 1965. (2 volumes.)

SILLEN, L.G. and MARTELL, A.E. *Stability constants of metal–ion complexes*. Chemical Society Special Publications 17 and 25. 1964 and 1971.

SMITHELLS, C.J. *Metals reference book*. 3rd edn. Butterworth, 1962.

STARK, J.G. and WALLACE, H.G. *Chemistry data book: 2nd edition in SI*. John Murray, 1982.

STEPHEN, H. and STEPHEN, T. *Solubilities of inorganic and organic compounds*. Pergamon, 1963–4. (2 volumes.) (Academy of Sciences of the USSR. Institute of Scientific Information.)

STULL, D.R. and SINKE, G.C. *Thermodynamic properties of the elements*. Advances in Chemistry series, volume 18. American Chemical Society, 1956.

SUTTON, L.E. *Tables of interatomic distances and configurations in molecules and ions*. Chemical Society Special Publications, 1958 and 1965 (supplement).

THERMODYNAMICS RESEARCH CENTRE *Selected values of properties of chemical compounds*, volume 1. Texas Agricultural and Mechanical University, USA, 1968.

UNITED STATES NATIONAL BUREAU OF STANDARDS *Electrochemical constants: proceedings of the NBS Semi-Centennial Symposium, 1951*. NBS Circular no. 524, 1953.

UNITED STATES NATIONAL BUREAU OF STANDARDS *Tables of chemical kinetics, homogeneous reactions*. Circular 510, 1951. Supplement no. 1, 1956. Supplement no. 2, 1960. Monograph no. 34, volumes 1 and 2, 1961 and 1964.

WAGMAN, D.D., EVANS, W.H., PARKER, V.B., HALOW, I., BAILEY, S.M., SCHUMM, R.H., CHURNEY, K.L., and NUTTALL, R.L. *Selected values of chemical thermodynamic properties*. Technical Notes Series 270-1–270-8. United States Department of Commerce, National Bureau of Standards, Washington DC, 1968–81.

WEAST, R.C. (ed.) *CRC handbook of chemistry and physics*. 62nd edn. CRC Press, 1981–2.

WELLS, A.F. *Structural inorganic chemistry*. 3rd edn. Oxford University Press, 1962.

WOOLMAN, J. and MOTTRAM, R.A. *The mechanical and physical properties of British Standard En steels*. British Iron and Steel Research Association, Steel User Section. Pergamon, 1964–6.

Substance index

a atomic size, mass, and abundance
ec electronic configuration
ie ionization energies
pt physical and thermochemical data.

a atomic size, mass, and abundance
ec electronic configuration
ie ionization energies
pt physical and thermochemical data.

a atomic size, mass, and abundance
ec electronic configuration
ie ionization energies
pt physical and thermochemical data.

---

a atomic size, mass, and abundance
ec electronic configuration
ie ionization energies
pt physical and thermochemical data.

General index

	I	II											III	IV	V	VI	VII	0
1s	1 H																	2 He
2s / 2p	3 Li	4 Be											5 B	6 C	7 N	8 O	9 F	10 Ne
3s / 3p	11 Na	12 Mg	**3d**										13 Al	14 Si	15 P	16 S	17 Cl	18 Ar
4s / 4p	19 K	20 Ca	21 Sc	22 Ti	23 V	24 Cr	25 Mn	26 Fe	27 Co	28 Ni	29 Cu	30 Zn	31 Ga	32 Ge	33 As	34 Se	35 Br	36 Kr
5s / 4d / 5p	37 Rb	38 Sr	39 Y	40 Zr	41 Nb	42 Mo	43 Tc	44 Ru	45 Rh	46 Pd	47 Ag	48 Cd	49 In	50 Sn	51 Sb	52 Te	53 I	54 Xe
6s / 5d / 6p	55 Cs	56 Ba	67 La	72 Hf	73 Ta	74 W	75 Re	76 Os	77 Ir	78 Pt	79 Au	80 Hg	81 Tl	82 Pb	83 Bi	84 Po	85 At	86 Rn
7s / 6d	87 Fr	88 Ra	89 Ac	104 Rf*	105 Ha	106												

4f	58 Ce	59 Pr	60 Nd	61 Pm	62 Sm	63 Eu	64 Gd	65 Tb	66 Dy	67 Ho	68 Er	69 Tm	70 Yb	71 Lu
5f	90 Th	91 Pa	92 U	93 Np	94 Pu	95 Am	96 Cm	97 Bk	98 Cf	99 Es	100 Fm	101 Md	102 No	103 Lr

*Or Ku.